Springer Praxis Books

More information about this series at http://www.springer.com/series/4097

Claudio Tuniz • Patrizia Tiberi Vipraio

From Apes to Cyborgs

New Perspectives on Human Evolution

 Springer

Published in association with
 Praxis Publishing
Chichester, UK

Claudio Tuniz
The "Abdus Salam"
International Centre
for Theoretical Physics (ICTP)
Trieste, Italy

Patrizia Tiberi Vipraio
University of Udine
Udine, Italy

SPRINGER-PRAXIS BOOKS IN POPULAR SCIENCE
Revised translation from the Italian language edition: La scimmia vestita. Dalle tribu di primati all'intelligenza artificiale by Claudio Tuniz, Patrizia Tiberi Vipraio,
© *Carocci editore 2018.*

Springer Praxis Books
ISSN 2626-6113 ISSN 2626-6121 (electronic)
Popular Science
ISBN 978-3-030-36521-9 ISBN 978-3-030-36522-6 (eBook)
https://doi.org/10.1007/978-3-030-36522-6

© Springer Nature Switzerland AG 2020

This work is subject to copyright. All rights are reserved by the Publisher, whether the whole or part of the material is concerned, specifically the rights of translation, reprinting, reuse of illustrations, recitation, broadcasting, reproduction on microfilms or in any other physical way, and transmission or information storage and retrieval, electronic adaptation, computer software, or by similar or dissimilar methodology now known or hereafter developed.
The use of general descriptive names, registered names, trademarks, service marks, etc. in this publication does not imply, even in the absence of a specific statement, that such names are exempt from the relevant protective laws and regulations and therefore free for general use.
The publisher, the authors, and the editors are safe to assume that the advice and information in this book are believed to be true and accurate at the date of publication. Neither the publisher nor the authors or the editors give a warranty, expressed or implied, with respect to the material contained herein or for any errors or omissions that may have been made. The publisher remains neutral with regard to jurisdictional claims in published maps and institutional affiliations.

This Springer imprint is published by the registered company Springer Nature Switzerland AG.
The registered company address is: Gewerbestrasse 11, 6330 Cham, Switzerland

Acknowledgements

This brief book is the result of many years of discussions with both experts and non-specialists about the many facets of human evolution and their relevance to imagining the future of *Homo sapiens*. We, therefore, owe a great deal to the many brilliant scholars upon whose research and ideas it draws. It is also an attempt to make recent discoveries accessible to all. We are grateful for the many comments of the general public and the high school and university students to whom these ideas were introduced. In addition, we would like to thank a number of friends and colleagues, who were extremely generous in commenting on the first draft and discussing it with us. Our sincere thanks go, in particular, to Fabio de Vincenzo and Davide Fiocco for their critical review of the manuscript and suggestions for improvements. We also thank Andrew Shuller for his help. Finally, we are deeply indebted to Cheryl Jones and Michaela Jarvis for carefully editing the manuscript in its various phases of development. Needless to say, any errors or inaccuracies that remain are all ours.

Introduction

There is currently unprecedented interest in understanding human origins. Several bestsellers, both fictional and non-fictional, have addressed some of the most pressing questions that have puzzled anthropologists for decades. How come a weak and naked ape became the master of the planet? Who are we and where do we come from? How do we relate with other species and among ourselves? How has the history of human evolution shaped our behaviour? How can we face the challenges ahead of us individually and collectively?

Drawing on a vast array of evidence, our storyline recalls the metamorphosis that affected the anatomy and behaviour of a lineage of primates that populated our planet in the last few million years. At one point, a self-reinforcing biological and cultural process helped develop an increasingly large brain in a number of human species. As *Homo sapiens*—or simply Sapiens[1]—we are the latest survivors of this evolutionary process. Many still maintain that our human species emerged as a result of continuous improvements—the so-called march of progress whereby we gradually evolved from bipedal apes into more beautiful and intelligent hominids. Most scholars now argue that this description of human evolution is deeply flawed. So, our story rejects the anthropocentric and linear discourse on humans' origins and their place in nature. Every living being emerges from a complex system and is conditioned by its environment. This often occurs through circular processes involving a network of networks and feedback loops; sometimes there are successful adaptations, other times extinctions. And the genus *Homo* is no exception.

This book provides a short sketch of these dynamic processes, focusing on the construction of human societies; it probes the life of our ancestors, starting from birth and extending through infancy, maternity, gender, diet, clothing, illnesses, labour, arts, entertainment and funerary rites. Three features seem unique to humans. First,

---

[1] For reasons of simplicity, in this book we will call "Sapiens" the members of the *Homo sapiens* species, "Neanderthals" the members of the *Homo neanderthalensis* species and "humans" the members of the genus *Homo*.

they show a strong tendency to extend their bodies and minds with tools and other cultural products (a sort of cognitive hybridisation). Though present in other animals, only in humans has this trait reached such unprecedented depth and scope. Second, they are able to generate imaginary worlds and enhance their theory of mind (understanding other people's thoughts) way beyond the capacity shown by all other species. This trait assists the formation of societies and differentiates them according to cultures. It provides stories and mental spaces that feel real and bounding. And finally, a last peculiarity of Sapiens, in particular, is the ability to connect in large numbers and generate complex and stratified societies according to predefined arbitrary rules, determined by history, geography, ethnicity, census, language and whatever is considered relevant to the purpose.

These capacities of Sapiens are explained by the specific traits—cognitive, anatomical, hormonal, genetic and neural—that emerged in the last 150,000 years. The new socioeconomic order was probably stimulated by a novel ecological niche, namely the availability of abundant, predictable and concentrated resources (mostly marine and lacustrine) that Sapiens encountered in Africa during the challenging harsher climatic conditions of the last glacial period. An embryonic H*omo oeconomicus* was probably conceived amid efforts to develop particular skills, problem-solving capabilities and institutions for exchange of goods and services.

The emergence of symbolic thought allowed for the formation of surpluses of goods and services, via the division of labour and the invention of money, long before farming and livestock breeding. In the late Pleistocene, this situation stimulated the first hierarchical societies, based on the authority and/or the credibility of the first leaders. Fear, trust and pleasure were nurtured and amplified through ceremonies based on storytelling, dance, music and art. The pro-social behaviour was based on a self-domestication process which is seen in animals where a reduction in natural aggression and an increase in communication are induced by extending a typical youthful behaviour to adulthood. Over time this process affected the anatomy, hormonal functions and genetic structure of some species. Self-domestication in humans, which accelerated during the last 50,000 years, contributed to the establishment of complex societies and the survival of our species. It is also significant, we will argue, to interpret our predisposition towards the future, increasingly intertwined with artificial intelligence (AI).

Throughout, this book engages rigorously with recent scholarship in a wide range of disciplines such as anthropology, archaeology, genetics, zoology, cognitive science and behavioural studies. It ends with rather more speculative sections on AI. In the Late Pleistocene, the human brain developed the capability to read, and perhaps, control thoughts and emotions in other individuals. Similarly, we are now instructing machines via digital algorithms to learn cumulatively and to read human emotions from facial expressions, to probe our tastes and feelings via social networks, search engines and online shopping sites. We are providing them with a working memory for reasoning and learning that imitates our own. We have domesticated other species with food and fear, but also with play and love. We have applied these tactics to some members of our own species (often successfully). Shall we be tamed, in turn, by the emotions and gratifications induced by means of intelligent machines? What are the costs and benefits of another—perhaps final—hybridisation with artificial intelligence? Are we going to be subject to a pervasive control of our social behaviour? Or is there a chance that—with the exception of a few adults—artificial intelligence will confine ever more infantilised humans to artificial worlds in which they can play and live happily ever after?

The book draws on the most recent ideas on human evolution, and its recent developments, stressing the importance of circular interactions between brain, culture and the environment, all mediated by body and objects. As a result, modern humans are able to build an extended mind that includes other people, cultural products and technologies. The book also highlights the entanglement of biology with culture, of genetics with global migrations, of economic drives with political gains and losses, to make inferences about our possible future. In doing so, it helps debunk ideas based on false information and prejudices about our central role in nature.

We will argue that the ability of Sapiens to form a social organism of increasing complexity was the secret of its success in the past, and still is, for the time being. However, that ability may venture into troubled waters beyond a certain level of complexity. This is when cooperation is outpaced by strong competition between humans for the exploitation of resources that are unevenly distributed, and when the degree of complexity is so high that it could be advisable to allow management by artificial intelligence.

These arguments are based on six main propositions.

1. Large and well-organised groups of Sapiens date back much earlier than previously thought. Different forms of embryonic societies emerged between 150,000 and 50,000 years ago, first in Africa and then in Eurasia. This point involves discussion of the latest archaeological discoveries and their synchronisation with climatic events.
2. Hybridisation, assimilation and replacement was common and widespread among all human species, as well as among different populations of Sapiens. This remark draws on the most recent results of paleo-genomics and very recent fossil discoveries and improved dating methods.
3. The combination of symbolic thought, complex language and culture together with self-domestication provides a robust explanation for the rise of large and composite hierarchical societies. This statement relates to recent historical and anthropological work and constitutes an extension into the fields of biology and political economy.
4. In particular, self-domestication helped reduce social conflict when assisted by specific cultural developments. It also led to reciprocal social arrangements and behavioural norms which constrained the freedom of individuals and affected power relationships, especially in regard to women. This observation extends work done in zoology and philosophy.
5. In endorsing the digital revolution and trying to apply AI to all our needs, we might overestimate our capacity to control it and risk losing the last degrees of freedom that self-domestication has not yet achieved. This is a suggested

interpretation of what may happen in the near future, according to our evolutionary trend.
6. By extending works from neuroscience, paleoneurology and cognitive archaeology we claim that our cyborg destiny was probably sealed in the Palaeolithic, when we started to incorporated tools and objects into our extended mind.

Though raising many questions, the book does not provide definite answers to all of them. Yet it does supply the reader with recent evidence and new hypotheses that should help him form an up-to-date opinion about the future. By illustrating the exploratory nature of scientific inquiry and its essential open-ended character, the book sets out to help the reader follow the various arguments together with how scientists have sought to approach them. It presents scientific complexity so that it may be understood with a minimum of hassle and does not presume any specific prior knowledge in an educated reader. Nevertheless, often thought-provoking, it does require an intelligent interest and an open mind.

Contents

1	**Our Deep History: A Short Overview**		1
	1.1	The Human Past	1
	1.2	Descended from the Apes?	6
	1.3	Our Place in Time and Space	9
2	**Human Biodiversity and Close Encounters**		11
	2.1	Sapiens: The African Origins	12
	2.2	Sapiens the Conqueror	14
	2.3	Neanderthals	17
	2.4	Denisovans	21
	2.5	Hobbits	22
	2.6	*Homo naledi* and *Homo luzonensis*	24
	2.7	Races and Racism	25
3	**The March for Hegemony**		27
	3.1	Tools, Fire and the Environment	27
	3.2	Once a Player, Always a Player	31
	3.3	Disappearance of the Neanderthals	32
	3.4	Man's Best Friend	34
	3.5	Self-Domestication of *Homo sapiens*	35
	3.6	Last Act	39
4	**The Naked Ape Dresses Up**		41
	4.1	The Naked Ape	42
	4.2	Coats, Shoes and Shelter	44
	4.3	Dress, Shame and Symbols	46
5	**The Evolution of Woman**		49
	5.1	Giving Birth: A Risky Business	50
	5.2	Monogamy or Polygamy?	52
	5.3	Taming the Female	54
	5.4	Sexual Selection: The Role of Women	57

6	**Work, Leisure and Learning**		59
	6.1	Growing Up Too Fast?	60
	6.2	Art and Entertainment	61
	6.3	Teaching and Learning	62
	6.4	Trust, Gossip and Shared Beliefs	64
	6.5	Work, Leisure and Learning Today	67
7	**Food for Body and Mind**		71
	7.1	Ritual Food	73
	7.2	Vegetarian or Carnivorous? Omnivorous	74
	7.3	Farmers and Breeders	75
	7.4	Us and Them	78
	7.5	Turning Our Genes On and Off	79
8	**Diseases and Grief**		83
	8.1	Disease and Therapies from the Past	84
	8.2	Diseases of the Present: A Possible Mismatch	87
	8.3	Funerary Rites	88
	8.4	The First Hierarchical Societies	90
	8.5	Bones, Tombs and Relics	91
9	**Brain and Mind**		95
	9.1	Brain and Mind in Deep Time	97
	9.2	Thinking of Thinking	99
	9.3	Modelling and Imagining	100
10	**Imaginary Worlds**		105
	10.1	New Realities	109
	10.2	Overcoming Perceptual Barriers	113
	10.3	Excess of Representation: The Economic Sphere	114
	10.4	Symmetric and Asymmetric Warfare	115
11	**Homo Oeconomicus**		119
	11.1	When It All Began	121
	11.2	What Is Money?	123
	11.3	Goods and Services	124
	11.4	The Private Accumulation of Wealth	127
	11.5	The Destruction of Collective Wealth	132
12	**Humans of the Future**		133
	12.1	Digital Networks	134
	12.2	Social Networks, Digital Games and Search Engines	137
	12.3	Market-Driven Digital Gurus	139
	12.4	Socio-Economic Networks	140
	12.5	Social Networks of Knowledge	141
	12.6	Global Networks and Territorial Networks	143
	12.7	Shrinking the Brain	145

12.8	Power and Pleasure		147
12.9	Truth and Post-truth		149
12.10	Intelligent Weapons and Preventive War		151
12.11	Transhumanism		153
12.12	Posthumanism		155
12.13	Deep Neural Networks		156
12.14	Tamed by Intelligent Machines		158

References 161

Chapter 1
Our Deep History: A Short Overview

Until recently, little was known about our deep past. History began with the first written records and the origin of our species was unclear. Human behaviour was put down to a mere biological drive of the primitive men and women. The main focus was on the emergence of human culture and the rise of great civilizations. Nor was it known how many and which human species had preceded our own. For millennia, we thought we were the only humans who ever existed. And on this misunderstanding, we have built narratives in which we were at the centre of the universe, the rulers of nature, the beloved of the forces of creation.

1.1 The Human Past

The discovery of the Neanderthals in 1856 was met with disbelief. Those who thought the species was a distant ancestor of ours were at first pilloried but later vindicated by the Neanderthal Genome Project. The project showed that everyone outside Africa shared a tiny proportion of their DNA with this extinct species. We co-existed for more than 200,000 years but lived in different continents most of the time. *Homo sapiens* was based in Africa, and *Homo Neanderthalensis* in Eurasia. Once the Sapiens went out of Africa, the two species lived next to each other for extended periods. We now know that many other human species existed before us, and others have been our contemporaries. Indeed, human evolution has often been misunderstood.

At first our species was said to emerge as a result of linear improvements in which we progressively transformed ourselves from bipedal primates into humans. Then, our evolution was portrayed by the metaphor of a tree whose branches represent different species. This image implies a sense of hierarchy and conveys a search for light—an aspiration to transcendence. A better metaphor is perhaps a coral reef in which the different twigs proliferate in all directions and flip open, like the different species of hominins, to occupy all possible ecological niches. The extinct hominins would then be analogous to all dead branches of a coral reef. Today, we know that all organisms evolve and adapt with much more complex mechanisms than those described by Charles Darwin in his original works. This consideration also applies to the subsequent "modern synthesis" which integrates the former theory with genetics. The genome is considered here as a sort of program that, from within the cell nucleus, plans the construction of living beings during adaptation to environmental conditions.

Yet even this last more modern scheme remains partial. It certainly functions in the framework of population genetics, but it does not explain other recently discovered phenomena. For example, the hybridization of species, including human species, is now well established, and needs to be considered. According to some, species are neither isolated branches of a tree of life nor of a coral reef. A better metaphor could be that of a braided river, a group of streams that weave into and out of each other, but are all part of the same system (Hawks 2016). These streams eventually merge into the sea (e.g. extant mankind).

We are getting closer now, but we are not there yet. In reality, evolution and adaptation take place through networks of continuous, non-linear interactions at different levels of the biological structure. These networks, we shall argue, link the different species to each other and to the environment through circular and retroactive processes. Both the nature and the dynamics of these processes deserve close attention. We will see that this new—multi-level—"darwinism of networks" is now supported by a consistent series of events that we keep discovering from the past and is associated with the emergence of complex social organisms and culture (Terradas 2017). The fabric of our narrative is made by interweaving biological and cultural evolution.

We will highlight the emergence of cognitive and social skills in an environment that was once characterized by high human biodiversity. We will discuss a recent hypothesis: that of a progressive self-domestication of our species, as an extension of what is observed in other species. Associated with words such as "taming," "breeding" and "guarding," domestication implies sharing the same "home" (*domus* in Latin), both literally and metaphorically.[1] Philosophers had discussed it extensively (Sloterdjik 2018). Recent zoological studies on domestication, performed on foxes and mice, show that domestication is the result of a biological syndrome that transforms anatomy, hormonal functions and genetic structure. The selection of less stressed and aggressive traits occurs with the maintenance of youthful characteristics in adulthood, slowing down the development process. Cortisol levels are reduced, while serotonin builds up. The growth of pro-social behavior occurs together with

---

[1]Charles Darwin was the first to systematically examine biological changes in species under artificial breeding conditions. Even though he did not refer to human self-domestication, in his studies on animals and plants.

He emphasised that their domestication is more than taming. It represents a goal-oriented process for human purposes. Eventually, three main features would distinguish domesticated animals from their wild ancestors: a higher variability of physical and cognitive capacity, a greater behavioural plasticity and educability, a smaller brain size (Darwin 1868).

more predisposition for sexual relations and play, all traits that are mediated by hormones and neurotransmitters of pleasure.

Dogs are the best example of the results of such a syndrome, which emerged as a process of self-domestication of some wolves during the last ice age. *Mutatis mutandis*, a self-domestication syndrome is said to be a robust hypothesis to justify the sharp rise in pro-social behavior of modern Sapiens. The direction of this great collective game is now increasingly entrusted to digital algorithms and artificial neural networks, which can be used to control and influence our behaviour. The possible consequences of this last development are discussed in the final part of this book.

After a very slow incubation period, that also involved some of our direct ancestors and was mainly anatomical, this process seems to have accelerated during the last 100,000 years. It is based on the selection of neural, genetic, physiological and behavioural innovations that transformed us into a social organism. And these circular mechanisms may shed light on a future increasingly influenced by artificial intelligence.

The extension of human history into the deep past rests on hard evidence: Written documents are not the only sources available. Other valuable information is written in our bodies and in the remains of our ancestors. Using the language of isotopes, teeth and bones of different hominids reveal when they lived and their environment. They also describe their movements across the landscape and even their diets. The microstructure of tooth enamel reveals the hominids' biological development and the stresses they endured when delivering children, for example (Smith 2018). We can reconstruct the external structure of their brains from their imprint on the inside surfaces of fossil skulls. From our current DNA and that of the humans who preceded us, we can shed light on past migrations and the degrees of kinship between the different human species.

There are also some epigenetic mechanisms[2], capable of switching gene expression on and off. These mechanisms respond to the environment in which ancient humans lived, changing the chemical conditions of DNA without changing the sequence of nucleotides in which the genetic code is written. Epigenetic influences are inheritable and can be identified in the bone remains, revealing diseases, hunger, and other stressful events suffered during life (Gokhman et al. 2017).

The analysis of brain activity when contemporary humans act out ancient behaviours opens another window on our past. For example, it can be shown that specific patterns of neural connections are activated during the production of certain Palaeolithic stone tools. It is also possible to identify the neurological and hormonal mechanisms that underlie our current behaviour, giving clues to how and why our species became such a complex social organism. The role of domestication in this process can be studied by comparing morphological and genomic differences between *Homo sapiens* and other human species that are now extinct

---

[2]The DNA nucleotide sequence takes many generations to change, providing information on long-term processes. In contrast, epigenetic effects occur quickly in response to environmental changes over the life of an individual.

(Theofanopoulou et al. 2017). Material products, such as tools and ornaments, help us to reconstruct cultures from the deep past. And science in now contributing a great deal to these efforts.

Before the new scientific methodologies could shine light on the deep past, archaeological finds unearthed by a small group of enthusiasts were studied with simple tools, such as callipers. These early scientists compiled catalogues and made comparisons. Thanks to their efforts, we know that men and women who preceded us gradually improved their stone tools. Some of these instruments have been traced back to the Palaeolithic period, which started in Africa about 2.5 million years ago. Others are more recent, from the Neolithic, which began in the Middle East 12,000 years ago, when stone was worked in a radically new way.

Eventually, we learned how to extract metals from stones and work them with fire and we started to breed animals and grow crops, revolutionizing our societies. Some of us accumulated a large amount of wealth, generating the first inequality. The scope for fighting wars widened. The rise and fall of the great civilizations, with rivalries and conflicts, soon became the backbone of history.

This simple plot thickens with the help of new technologies and methods, calling for a revision and an update. Recent discoveries suggest that some stone tools were made much earlier than previously thought—more than three million years ago.[3] Moreover, there is some evidence that the accumulation of wealth began long before the agricultural revolution. Abundant and concentrated resources, subject to proprietary exploitation, already existed. They were in many coastal areas of Africa and in the steppes of Europe and Asia, which were populated by large mammals, and represented a precious supply of raw materials, food and shelter during the late Pleistocene.[4] The difference in opulence of burials found in Eurasia during this period reflects the advent of hierarchical societies. Perforated shells, painted and finely worked, presumably enlaced to form bracelets and necklaces, were fashionable in parts of continental Africa as early as 80,000 or 100,000 years ago. They hint of nascent trade in the first organised societies.

Our larger frame of reference is the heterogeneous family of bipedal apes that separated from the chimpanzee line about seven million years ago. This evolutionary line embraces the genus *Homo*—which comprises about ten known species (Schwartz and Tattersall 2015) and a few more archaic species. Our account covers the transition from prey to predator as brain size increased. It ends with the onset of symbolic thought—i.e. the use of symbols to convey ideas and the capacity to reason in hypothetical terms—and the consolidation of cultures.

To understand the formation and dynamics of societies, we can rely on the findings of those late Palaeolithic[5] burials in Eurasia, which are sometimes rich in precious objects. They shed light on the dearly departed and on the economic and social relations of the time. Further information comes from the layout and

---

[3]That is before the beginning of the Pleistocene (which spans 2588 million to 11,700 years ago).
[4]This period extends between 126,000 and 11,700 years ago.
[5]This period spans, in Eurasia, 40,000–10,000 years ago.

architecture of settlements, from the materials used, even from garbage. The social sciences give us a glimpse of many aspects that might have passed unnoticed and offer insights into the lingering intractable mysteries of the remote past.

We will argue that what made us so different from other humans lies in our ability to imagine worlds different from those we live in. This ability was present in an embryonic state in other humans. But it was only in our species that it allowed us to take off and spin in novel trajectories of survival and growth. This can be attributed to the conjunction of a series of favourable circumstances and to self-reinforcing loops. At the dawn of humankind, we began to understand that we could satisfy our needs through the use of materials and tools that were not part of our bodies. Despite lacking some physical leverage of competitors, we could enrich ourselves through our work and ingenuity. This advantage, which distinguished humans from earlier hominins[6] was long limited to fire and tools that required minimal work. It was the product of small innovations to transform objects collected in nature. There was little progress for millions of years.

Technology was about to develop much more quickly with the emergence of symbolic thought. This new capacity blossomed in those humans we call "modern".[7] It enabled them to expand their societies, forfeiting individual independence in exchange for the possibility of enjoying more and more goods and services, via embryonic specialisation and exchange. Access to these was predicated on laws of co-existence and the enforcement of those laws. We now form societies of strangers. We generate complex cultures that inspire both individual and collective behaviour. We have become a social organism. Central to the development of human societies was a drive to control nature. Eventually we moved into an epoch, the so-called Anthropocene, in which we began to have a major impact on the global environment. This phase is often said to have started with the latest industrial revolution.[8] However, we contend that it dates back more than 50,000 years to the first extinctions of the megafauna, the large animals that once roamed the Earth, causing a major disruption of entire ecosystems. By those days, humans had walked the Earth for millions of years. They had departed from the common African ancestor—a still elusive figure—that we share with contemporary chimpanzees. When and how did the first humans appear?

---

[6]In reality, adoption of external tools useful for survival also is observed in other species, not only among primates but also in birds. But it was only in humans that it became so pervasive, systematic and aimed at all our needs.

[7]Our definition of "modern humans" refers to the cognitive capacities and enhanced pro-social behaviours of *Homo sapiens* that emerged in the last 100,000 years.

[8]Other scientists claim that the Anthropocene started in 1945, when we exploded the first atomic bomb, leaving an indelible signal in the geological history of the Earth, to which we are now adding many other climatic, biological and geochemical signs (Waters et al. 2016).

1.2 Descended from the Apes?

Setting the scene for the dawn of humans takes us back to the "Planet of the Apes" between 25 million and 15 million years ago. At the time, Africa and Eurasia were captives of the movement of their respective tectonic plates. A warm global climate promoted immense evergreen forests over the land from the equator to the poles. The hominoids (40 species from at least 30 genera have been discovered) spread over an area that ranged from the Iberian Peninsula to China and southern Africa. But things were about to change. About 50 million years ago, the Indian plate collided with the rest of Asia, throwing up the Himalayas and raising the plateau of Tibet. This altered atmospheric circulation and favoured the absorption of carbon dioxide in the newly formed rocks.

Starting then, the Earth's global temperature began falling. But this long-lived trend was interrupted, and reversed, between 25 and 15 million years ago. At the end of that warming phase, the global temperature started to fall again. Many areas became increasingly dry, splitting the habitats of the hominoids. This put an end to the planet of the apes. Some became extinct while others adapted to the new environments. Little remains of this ape biodiversity. In Asia, in addition to the gibbons, only two species of orangutan survive, in Borneo and Sumatra. In Africa, two species of gorillas, plus the bonobos and chimpanzees, cling to life. The last surviving human species has expanded on the whole planet. How did it happen? Here are a few hints.

We now know that walking upright would have been imperative to free the upper limbs and allow for the first use of instruments. The geological records reveal that about seven million years ago, when the planet of the apes was disappearing and the forests began to yield to the savannas, various bipedal creatures appeared in Africa. They were different from us and from other present-day apes. It is thought that the erect posture started randomly in some hominoid groups and was then imitated and passed on to subsequent generations because it was adaptive. It didn't necessarily take place in the savannas.

Chimpanzees have been observed spending time in the trees and on the ground at different times of the year, depending on the gradient in the microclimate of those settings. They are influenced by temperature, humidity and the availability of food at different heights. It has been suggested that the increase in seasonality and the prolongation of dry months during the late Miocene, between ten million and five million years ago, promoted the terrestrial life of the first hominins. This was when they still lived in the humid environment of some African forests (Takemoto 2017). Later, as the savannas expanded, this option turned into an evolutionary advantage.

Ardipithecus ramidus, a hominin that lived about 4.4 million years ago, was already bipedal (White et al. 2009). Consider the most famous specimen of this species, a female nicknamed Ardi. Although her new bone structure allowed her to walk upright, she retained long arms and prehensile feet with opposable toes to climb trees. She did not have, though, the modern specializations, such as knuckle walking, of the chimpanzee (White et al. 2015), which evolved along their own lineage. Lucy, a female of the species *Australopithecus afarensis*, appeared about a million years later (Walter 1994), in the same African regions. Her legs were longer, her toe was not opposable and her foot, rigid but already slightly arched, almost like ours. Nevertheless, Lucy's gait was probably still rather uncertain and staggering (Tobias 1997). At the time, there were many hominins of different species living a few kilometres apart in the Afar valley.

Time, space and purpose do not allow for a thorough account of all the human species that preceded our own. With respect to fossil records, humans are defined as associated with large brains, in relation to body size, smaller teeth, jaws and related musculature, relatively longer legs, the loss of climbing features, a particular shape in the body cage and an extended period of childhood. In East Africa and western Asia, the first fossil records consistent with such a pattern date from 1.5 to 1.9 million years and are generally assigned to *Homo ergaster/erectus*, which is believed to have evolved in Africa about two million years ago and then dispersed throughout Eurasia.

With respect to survival performance and behavioural traits, humans are also defined as those who first learned how to shape stone tools, as did *Homo habilis*, a species who lived in the previous million years, and perhaps some other species unknown to us. We shall provide more details about contemporary human biodiversity in the next chapter. Details will be included when we touch upon habits that different human species developed in the deep past.

Considering the extinction of all other human species, a brief assessment of the genetic differences between *Homo sapiens* and the chimps—our closest living relatives—can be instructive at this point. There are two species of chimp—*Pan troglodytes*, the common chimp, and *Pan paniscus*, the bonobo. The genome of the latter was sequenced in 2012 (Prüfer et al. 2012). Focusing on our genetic differences, we now know that we Sapiens share 1.7% of our DNA only with common chimps and 1.6% only with bonobos. What does this double connection suggest? And, more intriguingly, what does the absence of an overlap suggest?

The differences in social and sexual behaviour between common chimps and bonobos are quite well known. They both live in bands of about 50, but they are organized in very different ways. The chimps form hierarchical societies led by an alpha-male polygamist. They can be violent and territorial, especially towards other groups. The bonobos are organized in matriarchal societies and tend to be promiscuous and rather peaceful. They often use their sexuality, without much gender discrimination, to smooth over social tensions and show female cooperation to contain male aggressiveness. However, despite being fun-loving, bonobos are not all "peace and love." And even common chimps can show solidarity and affection. How did the speciation of bonobos generate?

According to recent studies (Takemoto et al. 2015), the separation of chimpanzees into two distinct species can be ascribed to the natural barrier generated by the Congo River. All ancestors of existing chimpanzees and bonobos had lived to the north of the river for millions of years. They competed with the ancestors of the gorilla. It seems that two million years ago, some ancestors of present-day bonobos managed to cross the river, populating a wide southern bend without the shadow of the gorilla. The different behaviour of bonobos compared with *Pan troglodytes* is explained by the existence of a resource-rich ecological niche free of competitors capable of posing a threat. Their less aggressive traits could be the result of self-domestication (Hare et al. 2012). On the north side of the river, however, competition with the ancestral gorillas selected common chimps for their territoriality and belligerence. To be true, the Congo River did not completely separate the two species. In fact, common chimpanzees share 1% of their genome with bonobos due to crossbreeding that occurred between 500,000 and 200,000 years ago, after the groups' separation. Even our closest living relatives have a history of ancestral crossbreeding similar to our own (de Manuel et al. 2016).

Homo sapiens has behavioural characteristics halfway between the two species of chimpanzee. In fact, for some traits, we Sapiens resemble them more than they resemble each other. However, our characteristics are very ambiguous and variable. It's as if we kept more degrees of freedom in our struggle for survival. As individuals, we can be very affectionate but also very aggressive. We are socially dichotomous—cooperative with like-minded people, such as family and friends, but highly competitive with various other groups. We have also developed a certain territorial ambiguity when it comes to settlement and dispersion. We attribute these characteristics to modern human culture and our difference from the rest of the animal world—a line which can't be crossed and which makes us unique. It is surprising that the genetic differences between us and our closest relatives are so small compared with cognitive and behavioural differences.

Recent studies tell us that these last differences are less significant than we thought. Even existing apes and monkeys exchange information through mutual observation (Hobaiter and Byrne 2014). They share food but only to eat in peace (Silk et al. 2013). They get excited watching TV (Kano and Hirata 2015), choose friends (Engelman and Herrmann 2016), know the sense of mourning (Yang et al. 2016), suffer from mid-life crises (Weiss et al. 2012), recognise themselves in the mirror (Gallup Jr. 1970) and use tokens with features similar to human money to exchange goods (De Petrillo et al. 2019). They also have a strong sense of fairness and can form societies based on collaboration and mutual support.

Recently, chimpanzees have been shown to be capable of learning the rules of a Chinese game called rock-paper-scissors. This game is difficult for those in hierarchical societies in which the transitive property of equality is valid (if Alpha is stronger than Beta, and Beta is stronger than me, then Alpha is stronger than me). But after a certain number of attempts, they learned the rules of circularity governing the game. They understood that if scissors trumped paper and paper trumped rock, it did not follow that scissors trumped rock (Gao et al. 2017).

The situation is hard to comprehend for those Sapiens who are more sensitive to domestication, as they are more prone to hierarchical relations in which the law of the strongest is bound to prevail. But the victory of the strongest is not guaranteed, for example, when the excessive use of force foments a wider opposition or when conditions are favourable for an asymmetric war, as we shall see later. Yet there are many differences between us and the apes. It is to this world of differences, but also of similarities, that we will dedicate some of the next chapters.

1.3 Our Place in Time and Space

To put our existence in perspective, and possibly argue against the paramount importance of humanity in the general order of things, a brief mention of where we stand in time and space may be useful to conclude this first overview. Going back in time, when did life appear on Earth? The answer lies in minerals in ancient Australian rocks. Traces of carbon of possible biological origin suggest that life appeared at least 4.1 billion years ago (Bell 2015). With the help of phylogenetics, we know that, ultimately, we all descend from LUCA, the Last Universal Common Ancestor, from which the first branches of life emerged. But what was LUCA? Its identikit was completed only recently.

According to the latest evidence, it was a single-celled organism, which had a very different lifestyle from ours because it loved the superheated depths of the Earth and lived in conditions like those found in current underwater hydrothermal sources. At the time, our planet was devoid of oxygen. But LUCA did not need it. Its food was prepared using what it found in that mephitic environment, rich in sulphur and iron. Its metabolism was based on hydrogen, nitrogen and carbon dioxide, which it transformed into ammonia and other organic compounds (Weiss et al. 2016). Its direct descendants lived in such conditions for billions of years, and many of them survive today in the depths of the Earth.

About two billion years later, some descendants of LUCA, the cyanobacteria, produced the oxygen necessary for the advent of eukaryotes,[9] i.e. the ascendants of all those multicellular organisms of which we are one of the latest expressions. But it was not until 630 million years ago that circumstances (a dramatic redistribution of nutrients in the seas) made them emerge (Brocks et al. 2017). The long journey of life that led to us followed a complex path through the animal kingdom, from vertebrates to mammals, primates and hominoids. It is at this point that our line—that of the hominins—diverged from that leading to chimpanzees and bonobos.

As we said, our species emerged between 300,000 and 200,000 years ago, in Africa, after a period when other ancient human species had populated different parts of the planet. From an anatomical point of view, our first ancestors were more robust and endowed with less reassuring features than those of most of our contemporaries. Their minds became equal to ours only from around 100,000 years ago with the emergence of increasingly complex cognitive and social traits. A few tens of thousands of years later, we Sapiens had already occupied several continents. In the meantime, every other human species went extinct. Our presence seems to have accelerated the reduction of biodiversity dramatically. It is possible to calculate how rich and varied the biodiversity of mammals would have been had we never existed (Faurby and Svenning 2015).

Our impact on the global biodiversity of vertebrates does not come only from hunting but, above all, from deforestation. A lesser-known but key impact is the

---

[9]Eukaryotes have complex cells, enclosed by a membrane containing the nucleus and mitochondria, both with a genetic code.

fragmentation of the forests with roads and other development. A recent study shows that these activities increase the danger of vertebrate extinction by 39% but favour the expansion of other species by 45%. Many of the favoured ones are invasive, so the net effect of human activity on biodiversity remains negative (Pfeifer et al. 2017). The benefits from our absence are proven by looking at the effects of the 1986 Chernobyl nuclear accident. The exclusion zone, which has remained uninhabited for more than 30 years due to radioactive contamination, has seen elk, deer and wild boar proliferate. The abundance of wolves has increased sevenfold (Deryabina et al. 2015).

In the past century, 177 species of mammals have died out. The sixth extinction, currently under way, has the characteristics of a biological annihilation (Ceballos et al. 2017), which consists of a progressive substitution of the flora and fauna inherited from the past by organisms designed by us. All this happened in a blink of an eye.

To get an idea of where we stand in time we could scale down the 13.8 billion years of existence of our universe to the six working days stated in Genesis for the creation of the world. Assuming that the Big Bang occurred at the beginning of a hypothetical Monday, the Earth would have formed the following Friday around 1 a.m., life would have appeared around 5 a.m. on the same day, the human line would have separated from that of the other anthropomorphic apes in the last 4 min of today (Saturday) and we, Sapiens, would have been born a few seconds ago (Tuniz 2012).

Chapter 2
Human Biodiversity and Close Encounters

Today, thanks to modern forensic methods, it is possible to reconstruct in every detail what humans and other hominids looked like. Perfect life-sized reconstructions of Neanderthals and australopithecines can be admired and even bought in the Elisabeth Daynès *atelier* in Paris. These realistic replicas speak of a history of human biodiversity that we should have suspected long ago from our observations of all other species. Our present uniqueness has led us to consider ourselves radically different. And on this basis, we have built the myth of our specialty on Earth, with a universe built around us, and all other living creatures at our disposal. Advanced scientific analyses of ancient human remains reveal many other details. They show, for example, that our species overlapped for some time with at least five other human species, and perhaps more. Human biodiversity was even greater further back in time. Yet we are the sole remaining representative of the genus *Homo*.

In the following, we will consider when and where *Homo sapiens* met its contemporary human species and what was the nature of these encounters, according to the most recent genetic discoveries. In general, genetic analysis sheds light on the difference between human species, the effects of their crossbreeding and details of their migrations. The mitochondrial DNA,[1] which is abundant and easy to analyse, is inherited only from the mother without any recombination.[2] Each branch of this genealogical tree is identified from "genetic markers"—DNA sequences that uniquely characterize a group of the population (called haplogroup).

Over time, mitochondrial DNA undergoes random mutations, or variations, in the nucleotide letters of the genetic alphabet, stretching along many generations. These mutations allow us to identify the different lines of descent. Moreover, the mutation

---

[1]The mitochondrial genome, with a circular double-helix structure, consists of 16,569 base pairs grouped into 37 genes. These bases are the four letters of the genetic alphabet A, G, C and T, denoting the compounds adenine, guanine, cytosine and thymine. The nuclear genome of the human cell has 3.2 billion pairs of bases, grouped in about 20,000 genes.

[2]We can also analyse the Y chromosome, which is handed down from father to son.

rate allows the calculation of the time elapsed since two human groups diverged from a common ancestor. However, the information inferred from mitochondrial DNA should be treated cautiously, as this is just a tiny portion of our full genetic code. More details on human migrations, population mixtures, expansions and replacements can be obtained from the analysis of the nuclear genome of present-day and ancient humans (Reich 2018). This also contains information on our crossbreeding with other human species.

2.1 Sapiens: The African Origins

Over thirty years ago, the DNA of today's Sapiens revealed some interesting news (Cann et al. 1987). Firstly, we could all claim a recent and exclusively African origin, although we are not sure how many waves of migrations took place out of this continent. Secondly, just before our Sapiens ancestors left Africa and spread throughout the planet around 60,000 years ago, they were very few in number and possibly on the verge of extinction. This can be inferred from the low variability of mitochondrial DNA in present-day populations, showing that we descend from the few survivors of a dramatic bottleneck. However, dispersing across the planet and increasing in numbers, our ancestors replaced all the other human species that had evolved from earlier migrants.

When and where did the first Sapiens emerge as species? Early genetic and archaeological data suggested that we evolved 200,000 years ago, in eastern Africa. This chronology was inferred from the mitochondrial DNA in living humans and confirmed by the age of the oldest Sapiens remains (Herto and Omo in Ethiopia, 165,000 years and 195,000 years, respectively).

Most scholars in human evolution studies embraced this recent Out of Africa model (Stringer and Andrews 1988), rejecting the assertion of the "multiregionalists" that *Homo sapiens* evolved simultaneously in different continents from the first humans who left Africa two million years ago (for example, *Homo erectus*). According to the latter theory, interbreeding with later migrants pushed our species along the same evolutionary line (Wolpoff and Caspari 1997). Multiregionalists are said to be "lumpers," as they claim that *Homo erectus, Homo heidelbergensis* and *Homo sapiens* were always the same species, following a linear evolution path. To them, the differences between today's Sapiens populations have been preserved for more than one million years. This has the controversial implication that racial differences originated in the deep past.

However, the narrative on Sapiens origins seems far from settled. Novel research suggests that the original version of the recent Out of Africa model has also been partially falsified, thanks to new fossil discoveries, the refinement of research on ancient DNA and improvements in dating techniques (Galway-Witham and Stringer 2018, Galway-Witham et al. 2019). The genomic analysis on South African human remains of 2000 years ago has allowed us to estimate the divergence of *Homo sapiens* from its archaic ancestors between 350,000 and 260,000 years ago

(Schlebush et al. 2017). This time span seemed to square with both the age of the *Homo sapiens* found in the cave of Florisbad, in South Africa—dated to about 260,000 years ago—and the archaic *Homo sapiens* remains found in Jibel Irhud, Morocco, and dated to 300,000 year ago (Hublin 2017; Richter et al. 2017).

The latter discovery would push the cradle of humanity back in time and, perhaps, move it to northwestern Africa. To be sure, the researchers suggest a Pan-African origin of *Homo sapiens*. At the time, the Sahara was green and allowed an ecological contiguity between northern and sub-Saharan Africa, as confirmed by the presence of a very similar fauna in the two regions. But even with this caveat, the discovery has already raised many doubts.

The skull from Morocco has some Sapiens characteristics, such as a small, vertical face. But it also has archaic traits, such as an elongated brain case. A globular skull instead has proven to be diagnostic of our species. These remains seem to reflect an important transitional phase for the emergence of our species in Africa, lending weight to the idea that our species appeared through a slow and gradual accretionary process (Neubauer et al. 2018). Yet this notion is far from being settled. Some scholars suggest that the Jibel Irhud skull belonged to *Homo heidelbergensis*, the presumed common ancestor of Sapiens, Neanderthal and Denisovans.

Many paleoanthropologists accept that we and other extinct human species share a common ancestor of African origin, which in Africa evolved into Sapiens and, in Eurasia, into Neanderthals. A possible candidate is *Homo heidelbergensis* (Manzi 2016). Another candidate is *Homo antecessor*, whose remains, dating back to about 850,000 years ago, were found in the 1990s in the Sierra of Atapuerca, Spain (Stringer 2016). His face shared more morphological features with Sapiens than with Neanderthals (Freidline et al. 2013). The same applies to some Chinese skulls, like Dali, dated to 300,000 years ago. If we could confirm that the Dali remains are Denisovans, e.g. with DNA analysis, we could infer that Sapiens and Denisovans had a face similar to that of the common ancestor, while Homo *heidelbergensis* and Neanderthal lost it (Stringer 2019). In other words, our face is not as "modern" as we think. Furthermore, the analyses of facial morphology suggest that *Homo heidelbergensis* is only the ancestor of Neanderthals (and possibly of other Eurasian humans) while *Homo sapiens* evolved from other *erectus*-like ancestors like Jibel Irhud hominin (Gunz et al. 2019).

In any case, it is confirmed that several human forms and species coexisted, both in Africa and in Eurasia, surviving the glacial and interglacial cycles that characterized the Pleistocene. Sometimes they were living in isolation, other times they had opportunities for overlapping and crossbreeding. During these extreme environmental fluctuations, human migrations can be studied using ecological approaches based on the dynamics between "source" and "sink" populations. In the former, a surplus of individuals is formed, favored by a greater availability of resources (Dennell et al. 2010; Martinon-Torres et al. 2018). In the latter, the scarcity of resources lowers the birth rate and reduces the population. During the glacial periods, the arid and cold steppes of Central Eurasia became uninhabitable, turning into sinks for the human populations, while the more southern areas retained woodland and forests, offering shelters suitable for survival. They would thus offer the source for the encounter of different human species and forms. Their interbreeding, when fertile, could have led to new branches of different human forms and species.

Under favorable climatic conditions, such as those established around 400,000 years ago, an evolutionary branch leading to Neanderthals, Denisovans and other human species could have developed and populated the boreal and deciduous forests of northern and central Eurasia. Fossils discovered in China—such as the Dali remains—present a mix of archaic and derived traits. They could be another example of these dynamics. But the pattern is unclear. Some argue that they are transitional forms between the Asian *Homo erectus* and *Homo sapiens* (Athreya and Wu 2017). Others maintain that they belong to African *Homo heidelbergensis*. Other scholars suggest that a human branch emerged from one of these sources (perhaps in the Middle East) and reached Africa, where it later evolved into Sapiens (Douglas 2018). In summary, we don't know much about our common ancestor with other human species, but likely he had a face similar to *Homo antecessor*, and his original home could have been Africa, Europe or Asia (Stringer 2019).

Most scholars believe that these ideas are excessively speculative and that more data and fossils are needed to develop credible theories. However, a growing number of paleoanthropologists agree that the model on our recent African origins needs to be updated. It should consider the complex interactions of hybridization and assimilation that have taken place between the different human species and between the Sapiens populations that lived in Africa and in Eurasia in different environmental niches. They are documented both by their fossil remains and by the remains of their material culture (Scerri et al. 2018).

Thanks to the contributions of genetics, physics and other "hard" sciences, paleoanthropology is now increasingly based on a quantitative approach. For example, X-ray microtomography can provide valuable information that allows, together with paleogenomics, more precise taxonomic and phylogenetic classifications. When applied to all available fossil teeth of the different Eurasian hominins, these analyses will allow us to shed light on the evolutionary story behind their distribution in space and time (Martinon-Torres et al. 2018).

2.2 Sapiens the Conqueror

Let's now consider what happened to the small groups of Sapiens that survived the bottleneck between 100,000 and 50,000 years before the present. Some of them certainly remained in Africa, where the evidence of new behaviour emerges as early as 80,000 years ago. What routes did our ancestors take in their conquest of the Earth? Some genetic evidence from contemporary Sapiens confirms that modern humans had already left Africa by 75,000 years ago (Pagani et al. 2016). Other evidence, both genetic and archaeological, suggests that their dispersal from Africa to Eurasia began earlier than the date assumed in previous models (Bae et al. 2017).

A jawbone found in the cave of Misliya, in current Israel, suggests that the Sapiens were already there 180,000 to 200,000 years ago (Hershkovitz et al.

2018). It also seems they appeared in China more than 100,000 years ago (Liu et al. 2010, 2015). A new study of a skull fragment discovered 40 years ago inside Apidina Cave, in southern Greece, suggests that it belonged to a Sapiens who lived in that area about 210,000 years ago (Harvati et al. 2019). In Indonesia, there are Sapiens fossil remains dating back to 70,000 years ago (Westaway et al. 2017), while in Australia their oldest archaeological remains have an age of 65,000 years (Clarkson et al. 2017).

The first colonizers of Australia had developed a very elaborate lithic technology, including polished axes. They also used mortars to pulverize minerals, such as ochre and mica, to make pigments, and tools to grind seeds. It is thought that this culture could only have been generated and supported by quite a big social group. According to recent calculations, the first colonization of the continent can be attributed to about 3000 individuals.[3] It was therefore a well-organized social group, and certainly not a "family on a raft drifting" (Williams 2013).

The analysis of nuclear DNA from an Aboriginal Australian revealed two waves of Sapiens, from west to the east (Rasmussen et al. 2011). The first, which began between 75,000 and 62,000 years ago in Africa, ended between 50,000 and 60,000 years ago on the former continent of Sahul.[4] The second brought the African Sapiens to Asia around 30,000 years ago, first populating the current China and then all of eastern Asia.[5] This second wave of people continued their journey to the Americas, crossing the current Bering Strait, which at the time was emerged land called Beringia. Beringia connected Siberia to Alaska. The genetic analysis of current Native Americans has revealed that there was indeed interbreeding about 20,000 years ago amid these two waves of migration through Asia before the migrants' descendants could reach the so-called New World.

The possibility of low-cost sequencing of the DNA of extant and extinct humans[6] opens the way to finding other details on when and where our migrations took place. It turns out that nobody can trace their origins to a specific territory. Those who settled were destined to meet and clash with passing populations, producing descendants, some of whom stayed while others continued to roam. Each region can tell us many stories of our ancestors from the comparison of the DNA of ancient human bones with that of our contemporaries. Consider, for example, Lake Baikal, in Siberia. A comparison of DNA from the 24,000-year-old remains of an individual found there with present-day Native Americans revealed that some of the latter were

---

[3]The assessment starts from the size of the Aboriginal population in 1788, estimated at between 770,000 and 1.2 million. The rates of population growth are calculated using the number of Aboriginal archaeological sites present in different eras, before the arrival of the English colonizers.
[4]That, at the time, was composed of the present-day Australian continent with neighbouring areas, including the present-day Papua New Guinea and Tasmania.
[5]This migration would have given rise to modern Asian people.
[6]At the time of this writing, National Geographic was offering a kit at the discounted price of US $149.95 for the personalized analysis of your DNA as part of the Geno 2.0 project. After receiving your sample, the company agrees to send detailed information on your ancestors back in deep times, on their migrations and on their interbreeding with other human species.

descendants of the Mal'ta, a group that lived near the Siberian lake between 24,000 and 11,000 years ago (Raghavan 2014). This lineage was confirmed by the genomic analysis of a child who lived in current Montana 13,000 years ago (Rasmussen et al. 2014).

But some Native Americans also have a European genetic component. Where does this come from? One would think from the recent European colonization of that continent. But it is not so. Comparing the mitochondrial genome of the Siberian individual mentioned above with that of the European populations of the Upper Palaeolithic, we see that he also has a direct relationship to the latter (Raghavan 2014). It follows that when the DNA of current Native Americans reveals their European ancestry, this is not only due to what happened after the rediscovery of America by Christopher Columbus. It also depends on what happened earlier in Eurasia. The idea that humans can be considered natives of this or that region, and that they have been able to develop in isolation makes little sense. In the long run, isolation is only a transitory phase. The history of all human migrations would take us out of the scope of this book. The most lasting isolation took place in Australia. A recent analysis of mitochondrial DNA from the hair of many present-day Australian Aborigines from various areas confirms what was said above: Their ancestors colonized Australia in a single wave, dispersing over a few millennia. Australia was their continent for more than 50,000 years (Tobler et al. 2017). No other people can boast this record.

In general, we Sapiens dispersed for thousands of generations. We changed our appearance to suit the environment and also due to genetic drift—genetic changes caused by random factors. Take skin colour, for example. Natural selection in latitudes of high solar radiation favours those with darker skin (Jablonski et al. 2010). But at latitudes of low solar radiation, darker skin inhibits the production of vitamin D, essential for the formation of bone and muscle tissue. Individuals with lighter skin would therefore be selected for.

This process, which is reversible through the generations, does not have to be immediate or rapid. For example, it took a long time for our species to become fairer during our first colonization of Europe around 45,000 years ago. The genetic modifications that reduced the pigmentation of our skin took place between 19,000 and 11,000 years ago (Beleza et al. 2012). This extraordinary fact was explained by presuming that the speed of adaptation to environmental conditions was positively related to the increase in population. In the early days, the Sapiens who went to Europe had remained quite contained due to bad climate. There was little scope for accelerated adaptation. Later, the archaeological register shows an increase in population that probably allowed a rapid selection of genes associated with lighter skin. This would explain the long incubation time of an adaptation that one would have thought would be quicker.

The recent reconstruction, based on DNA, of the appearance of Cheddar Man, who lived in England 10,000 years ago, adds another piece to this puzzle. Among other bewildering facts, it was discovered that, despite having blue eyes, his skin was very dark and he had curly hair (Brace et al. 2018). This suggests that the emergence of a fairer complexion could be delayed if the diet continued to be very rich in vitamin D, like that of hunter-gatherers. It would have been the advent of agriculture and a grain-based diet poor in this vitamin to provoke the spread in Europe of a population with lighter skin.

So far, we have covered some aspects of our planetary peregrinations as Sapiens. What of our meetings or clashes with other human species?

2.3 Neanderthals

When Sapiens arrived in Europe, the Neanderthals had already occupied a vast territory, from the Mediterranean to Siberia, for hundreds of thousands of years. They had survived various glaciations unscathed.

Neanderthals have long been described as the primitive ancestors of we pale and civilized Europeans. Nothing could be further from the truth. They were not our ancestors, they were not particularly primitive and we were not yet pale. But we survived and they became extinct. Why? We don't know for sure, but if we combine all the information that has emerged in recent years, we can make better conjectures than those of the past. Let's begin by making an identikit. The Neanderthals were very robust with a stocky body and funnel-shaped rib cage. The forehead was sloped down towards protruding eyebrows, a large nose and wide eye sockets. The skull, flattened and stretched backwards, had a chignon shape. In recent times we have discovered other details. The sequencing of their DNA allowed us to establish, for example, that they could have had red hair, and clear, freckled skin[7] (Lalueza-Fox et al. 2007).

If a suitably dressed Neanderthal walked down a crowded street today, it could try to pass unnoticed, but it would hardly succeed and indeed create some apprehension. Our direct African ancestors were physically very different from the Neanderthals, due to the climate and environment. They were more agile and slender, with the skull less stretched back, the nape and face more vertical, a high forehead and a very pronounced chin. But they were not very similar to contemporary Sapiens. They could perhaps resemble a very rough version of some of us—those who eschew the softness of "civilization" and who are so celebrated in the Hollywood mythology of the macho man. In reality, we began to look more gracile and infantile from the very start. Gradually we developed a smaller face, more modest eyebrows, and a much more globular skull—a characteristic typical of younger individuals—that we don't share with any other human species that has ever existed.

This process, called neoteny (looking like children), has accelerated over the past 40,000 years. It is said to form part of a domestication process, i.e. a survival strategy

---

[7]They had a variant of the MC1R gene, which promotes pale skin and red hair. Another gene causes the same effect in Sapiens.

aimed at reducing aggression during the formation of our social groups.[8] We will discuss the implications of this trend later. For now, it will do to note that, according to some paleo-neurologists, it was precisely that roundish skull that made the difference in Sapiens' struggle for survival. The skull had expanded first laterally in previous humans and then—only in our species—in its upper surface, creating space for the development of brain areas that would allow for the increase of our cognitive abilities (Bruner and Iriki 2016, Bruner et al. 2017; Bruner 2018; Bruner 2019).

In contrast, the evolution of Neanderthal skulls favoured the development of areas that improved eyesight, a winning feature in conditions of less solar radiation such as those typical of Eurasia (Pearce et al. 2013). These comparisons must be treated with caution, however. Although we can make approximate evaluations of the external brain structures, we cannot study the brain functions comparatively. However, we can deduce Neanderthal behaviour from the objects they could make. It is only in this sense that we can hazard a guess on their cognitive abilities.

Neanderthal lithic tools, called Mousterian, have been studied for more than a century. If we focus on the latest findings, however, surprising cultural aspects emerge. In the Fumane cave near Verona, Italy, it was discovered that the Neanderthals adorned themselves with bird feathers, plucked out with very sharp lithic tools. Bones there belong to the monk vulture, the cuckoo hawk and other large birds of prey with particularly leathery flesh. It is likely that the birds were hunted only for their beautiful coloured feathers (Peresani et al. 2011). In some Neanderthal caves of northern Italy and Croatia, eagle claws were found, modified to be used as personal ornaments. One can still see the signs of the stone tools used to extract them from the legs of these birds of prey (Romandini et al. 2014; Radovčić et al. 2015).

It was argued that the Neanderthals, like our ancestors, painted their bodies and adorned themselves with perforated shells (Zilhão et al. 2010). This behaviour has been confirmed by the recent discovery of perforated shells bearing traces of ochre pigment in the Spanish cave of Cueva de los Aviones, with uranium/thorium dates of 115,000–120,000 years (Hoffmann et al. 2018a). A century ago, the Neanderthals had a brutal and menacing image. Now we know their appearance could have been much more captivating and "new age." But we don't know whether self-decoration was gender-specific. There is also evidence for Neanderthal artistic acumen. Examples include 41,000-year-old cave paintings from the cave of El Castillo, in northern Spain. Some were attributed to the Neanderthals (Pike et al. 2012). In the cave of Gorham, in Gibraltar, elaborate engravings present abstract motifs, which have been assigned an age of 39,000 years (Rodríguez-Vidal et al. 2014).

Their Neanderthal provenance is confirmed by Mousterian stone tools in associated cave sediments. Extraordinary rock paintings, including hand-stencils, were

---

[8] According to recent studies (Gokhman et al. 2017), this transformation would be due to epigenetic effects, which would have modified our vocal tract (larynx, vocal cords, epiglottis, hyoid bone and tongue) after our separation from the evolutionary line that led to the Neanderthals and the Denisovans.

recently found in the Spanish caves of La Pasiega, Maltravieso and Ardales. They date back more than 20,000 years before the arrival of modern humans in Europe (Hoffmann et al. 2018b). Neanderthal handiwork was also utilitarian. Bone tools, so-called *lissoirs*, from a French site attributed to Neanderthals, were used to waterproof the animal skins they wore. We Sapiens would later copy them (Soressi et al. 2013). They used their teeth to work the material, judging from the wear and tear on incisors from many Neanderthal skeletons.

The enamel of the anterior teeth of a 50,000-year-old Neanderthal skeleton from La Ferrassie, France was so eroded that the dentine and pulp chamber were exposed. The use of the teeth as a tool seems to derive from poor development of those components of the parietal lobes that preside over the coordination between visual and manual functions during the execution of complex processes (Bruner et al. 2014). The teeth could have been a kind of "third hand" to help Neanderthals perform some complex tasks. Studies of about 100 incisors and canines from European Neanderthals reveal a division of labour on gender grounds (Estairrich and Rosas 2015).

Our "cousins" might also have had complex language. That's the conclusion of an analysis of a Neanderthal hyoid bone from the cave of Kebara, in Israel, dating back 60,000 years (D'anastasio et al. 2013). The only bone in the vocal tract, the hyoid is the only part of that anatomy that can be fossilized. The microstructure of the hyoid from the Neanderthal of the Kebara cave is similar to that of modern humans. Moreover, in all the samples compared, the histological detail is typical of a bone subjected to intense and continuous metabolic activity. The comparison between the Neanderthal and Sapiens hyoids shows significant similarities in their biomechanical behaviour. We can therefore say that their hyoid bone was used regularly to emit sounds, just like ours. But we do not know how complex their language was.

What happened when Sapiens and Neanderthals came into contact? We can certainly imagine opportunities for conflict, but no evidence remains. However, we have evidence on the nature of some of those meetings. It emerges from the most recent genetic studies. In 2010, the genome was sequenced from the remains of a Neanderthal femur from the cave of Vindija in Croatia, dating back 38,000 years (Green et al. 2010). It was not easy to get. The DNA was degraded, due to the action of bacteria and fungi, and shattered into no more than 200 base pairs. Fortunately, modern genetic sequencing techniques allow the reconstruction of DNA's original structure, even in fairly small samples. The results showed that a small part (1–4%) of the Neanderthal genome was present in the genomes of all Europeans and Asians today. It was absent from Africans. It follows that our meetings with this species would not always have been hostile.

It is difficult to establish the degree of consent that would have characterized all these relationships. But we cannot rule out the possibility that the two species could have been on reasonably good terms. When would these meetings have happened? Surely after our dispersal from Africa, which probably began earlier than 100,000 years ago. Some Sapiens groups might have met the Neanderthals in the territory that is part of the present Middle East. The periodic presence of both species in the region is confirmed by the fossil remains found in the caves of Qafzeh and Skhul, near Nazareth.

During the last interglacial period, between 130,000 and 110,000 years ago, the Sapiens were dispersing northwards along the entire African continent. The Sahara was much more pleasant (Timmermann and Friedrich 2016; Tierney et al. 2017). The isotopic analyses of plant residues carried by the wind and deposited in the sediments of the Atlantic Ocean, reveal lakes, streams and rich vegetation where today there are only sand dunes (Castañeda et al. 2009). Toward the end of that distant period at the beginning of a new ice age, various Neanderthal groups were abandoning Eurasia to seek refuge in the milder lands of the present Middle East. It is here that some fond encounters probably took place. But we have learned that only due to traces found far from those places. Believe it or not, evidence for these meetings was found in a remote region of Siberia.

The analysis of the genome of a Neanderthal woman recently discovered in the Altai mountains, where she lived 50,000 years ago, reveals that her ancestors met the Sapiens about 100,000 years ago (Kuhlwilm et al. 2016). Other such encounters might have taken place earlier. As mentioned before, there is growing evidence that Sapiens were already present in some Eurasian regions occupied by Neanderthals as far back as 200,000 years ago. Lacking a modern mind, those Sapiens built tools similar to those used by the Neanderthals, for example those attributed to the so-called Levallois technology.

It seems that the intimate encounters between the two species did not occur often, but only when the environmental conditions permitted. Thousands of years might have elapsed between one fertile meeting and another. In any case, our contiguity with the Neanderthals endured. This continuity was confirmed by the recent discovery of a 55,000-year-old Sapiens skull in present-day Israel. The area was probably still frequented by Neanderthals in those days (Hershkovitz et al. 2018). Gene exchanges occurred in the Middle East and Europe until the extinction of the Neanderthals. A recent DNA analysis of the remains of a Sapiens of 40,000 years ago (discovered in Romania) reveals that he had a Neanderthal great-great-grandfather (Fu et al. 2015).

The dental plaque of a 50,000-year-old Neanderthal found in present-day Spain provides more "soft" details of the nature of these relationships. The analysis of the DNA of the organic material contained in it shows the presence of the bacterium, *Methanobrevibacter oralis*, which is also in our mouths. The genetic comparison of our bacterium and that of the Neanderthal in question reveals that the two bacteria separated about 100,000 years ago. As a result, Sapiens' *M. oralis* cannot be traced back to the common ancestor. Something must have happened after our split into two evolutionary lines. It has been suggested that the bacteria were transmitted amid some act of affection (Weyrich et al. 2017). "If you're swapping spit between species, there's kissing going on, or at least food sharing, which would suggest that these interactions were much friendlier and much more intimate than anybody ever possibly imagined," said Laura Weyrich (quoted in Callaway 2017, p.163). However, an occasional exchange of saliva might have occurred in many not-socosy circumstances. This minor contention is reported to suggest that it would be wise to exercise a little caution when we tell tales of Pleistocene sex and love.

Not all geneticists agree that the Neanderthal component of our genome is necessarily due to hybridization. It is theoretically possible to attribute the component to the genetic characteristics of the ancient geographical dispersion of humans of African origin before their departure from that continent (Barbujani and Colonna 2010). We even have a model consistent with the absence of hybridization. It is obtained by simulating the dispersal (Eriksson and Manica 2012). This thesis finds limited support at the moment. We endorse the prevailing interpretation.

We now turn our attention to the relationship between modern humans and the two other human species with whom we shared the planet for a while. We cannot say much, for the evidence is much scarcer and only recently compiled. Nevertheless it is interesting to mention a few facts and some hypotheses.

2.4 Denisovans

Researchers at the Max Planck Institute in Leipzig, Germany, published the genome of a third human species whose remains were first found in the Siberian cave of Denisova. The cave also bears the mark of Neanderthals and Sapiens (Krause et al. 2010; Reich et al. 2010; Meyer et al. 2012). It has hosted three human species in overlapping periods. The only fossil remains found in the cave and attributed to this new species are the phalanx of a finger and three teeth, dating back to between 30,000 and more than 100,000 years ago. A 160,000-year-old piece of mandible was recently found on the Tibetan Plateau and attributed to the same hominin using ancient protein analysis (Chen et al. 2019). It is amazing how much information can be derived from so few remains, thanks to the progress of science and the intersection of the different data available. For example, new methods based on the chemical modification (methylation) of DNA in fossil fragments can predict the anatomical morphology of the Denisovans. It seems they shared with the Neanderthals an elongated face and a wide pelvis, but had a more pronounced dental arch and lateral cranial expansion (Gokhman et al. 2019).

The genetic analysis shows that the Denisovans occupied a vast region extending from Siberia to Oceania. They overlapped with the Neanderthals and our Sapiens ancestors. Traces of these meetings are recorded in the DNA of some of us. It corresponds to about 4–6% of Denisovan DNA. It is fairly high in the Sapiens who have settled in Tibet and in present-day Oceania. When and where did these meetings take place? The answer is currently based on circumstantial evidence.

The best hypothesis is that Sapiens and Denisovans met in Eurasia during the dispersions of the former group eastward between more than 100,000 and 60,000 years ago. A very recent discovery, however, aroused suspicions that such meetings took place elsewhere. Such arguments cast doubt on the idea that Sapiens were the first to cross the Wallace Line, a theoretical boundary that has always separated the Asiatic fauna and flora from Sahul. It existed even during the last glacial period when the sea level was more than 100 meters lower than today. However, hundreds of stone tools dating back 120,000 years have been found in Sulawesi, an island east of the Wallace Line (van den Bergh et al. 2016a, b). Who made them?

Since no human remains from this period have been discovered on the site, we don't know. The lithic tools could be attributed to at least three species: to *Homo floresiensis* (which lived on the neighbouring island of Flores as long as 190,000 years ago), to *Homo erectus* (present in neighbouring Java as long as 1.5 million years ago) or to the Denisovans (if we rely on genetic information present in the current indigenous people of Papua and Australia). If the presence of the Denisovans in current Indonesia were confirmed, the species would have crossed the Wallace Line tens of thousands of years before Sapiens, and then met us at the end of our dispersal from Africa to Sahul.

This paleo-gossip is far from prurient. The hybridization of the different human species has profound implications for the health of living people. We will cover the negative effects of our intersection with the Neanderthals later on. We can anticipate some positive effects. Some Neanderthal genes have helped us adapt to the cold climate and the low levels of solar UV radiation in northern Eurasia. The variant of a gene called EPAS1, inherited from the Denisovans, was useful to the Sapiens who settled in Tibet (Huerta-Sánchez 2014). Since it favours metabolism in a low-oxygen environment, it has facilitated their adaptation to altitudes above 4000 m. This gene no longer exists in any human population. The above-mentioned discovery – in Tibet – suggests the Denisovans were able to adapt to high altitudes long before the arrival of Sapiens. Even the future inhabitants of Sahul had met the Denisovans. The analysis of the hair taken from an Australian Aborigine by a British ethnologist in 1920 shows that he (or she) had both Neanderthal and Denisovan heritage. But there is no EPAS1. Perhaps it was no longer needed in the Australian plains, and was lost by natural selection (Rasmussen et al. 2011).

2.5 Hobbits

In 2003 in the wondrous cave of Liang Bua on Flores, the remains of some specimens of a fourth bizarre human species were found. The remains were classified as *Homo floresiensis* (Brown et al. 2004; Morwood et al. 2004, 2005). The species was nicknamed Hobbit. It seemed impossible that human beings could exist with such strange morphology. They were less than one metre tall, and had huge flat feet, long arms and small brains. It seems, however, that they made stone tools with which they hunted large and dangerous animals. They fed on gigantic rats, on a species of small elephants that are now extinct, and even on huge monitor lizards that survive on Flores and the nearby islands of Rinca and Komodo.

Immediately after the discovery of the Hobbits, it was hard to believe that they belonged to a new human species. The remains were suspected to belong to some deformed Sapiens, affected by microcephaly, a malformation in which the brain remains very small and the individual very short.[9] Others claimed that the remains belonged to a Sapiens affected by Down syndrome, a disease that would cause brain shrinking, shortening of the limbs and flattening of the feet (Henneberg et al. 2014).

The scientific community continues to reject these interpretations (Baab 2012, Baab et al. 2016). Other paleoanthropologists suggest that they evolved from some archaic man, like *Homo erectus*, and were dwarfed by living on an island where resources are scarce.[10] Other scholars assume that Hobbit was a variant of *Australopithecus* (Baab et al. 2016). Still others, using phylogenetic analyses,[11] suggest that Hobbit's ancestors were early representatives of the genus *Homo* (Argue et al. 2017). More remains attributed to *Homo floresiensis* were found at another site in Central Flores, Mata Menge, with ages of 700,000 years (van den Bergh et al. 2016a, b), supporting the hypothesis that the small-bodied hominins might have been endemic to the island.

The Hobbit's DNA has not been sequenced, as of yet. Therefore, we cannot say anything about whether, when, and with whom, they might have interbred. However, some evidence is now available about the times in which the Hobbits disappeared, and their possible encounters with Sapiens. The first dates (radiocarbon and luminescence) suggested their survival until 17,000 years ago. But now we have more reliable dates, obtained with precise uranium geo-chronometers. They show that the species disappeared from the archaeological record 60,000 years ago, coinciding with the date of Sapiens arrival in the region (Sutikna et al. 2016).

This throws a sinister light on the expansion of modern humans eastward. We now know that we were not the only humans to disembark in current Indonesia at the time. And the disappearance of the Hobbits could have been the result of a feud between different human species amid climate catastrophes. As Richard Roberts, one of the Australian scientists involved in the new dating said: "We found the smoking gun of that interaction, but not the bullet." Yet maybe the mystery is now solved (Callaway 2016). Two molars of Sapiens were discovered in Liang Bua. Evidence is mounting that our species wiped out the Hobbits as well as other human species. Prudence suggests waiting for more evidence to support this claim.

---

[9]This is a fairly common academic reaction. After the discovery of the first Neanderthal, in 1856, the species was thought to be a Sapiens with rickets.

[10]This phenomenon, typical of islands, left the fossil remains dating from the ice ages of dwarfed elephants in Sicily and dwarfed mammoths in Sardinia.

[11]The phylogenetic reconstruction is based on the analysis of traits that allow the assessment of the most probable evolutionary relationships among hominins. In this study, the features considered were the structure of the skull, jaws, teeth and limbs.

2.6 *Homo naledi* and *Homo luzonensis*

The Hobbit was not the only human species that survived until Sapiens came on the scene. Recently, another archaic species was identified in South Africa, *Homo naledi*, who lived until about 200,000 years ago. Its discovery was serendipitous. The remains of more than 15 individuals were found in a complex system of underground caverns near Johannesburg. Reaching them required crawling inside narrow, tortuous tunnels.[12]

Like *Homo floresiensis*, *Homo naledi* would have looked odd (Berger et al. 2017; Hawks et al. 2017; Dirks et al. 2017). Along with very archaic characteristics— arboreal, curved fingers and a pelvic structure typical of hominids living two million years ago—it had teeth, jaws, wrists, palms, feet and legs like those of more recent humans. Its brain differed structurally from that of *Homo floresiensis*, and, at more than 600 mL, was a little bigger. Recent studies also show some human-like structures in the external surface of its cortex (Holloway et al. 2018). Despite its small head, *Homo naledi* was almost as tall as modern humans. Since no animal remains or tools have been found in the cave, little can be said about its behaviour. Nevertheless, the discoverers hypothesise that the stone tools and use of fire, documented in other African archaeological sites of the same age, could be traced back to this species at a time from which there are no other hominin remains.

Also for this hominin, there is no genetic analysis available and therefore no hints whether some members of our species carry a tiny fraction of *Homo floresiensis* or *Homo naledi* DNA. The Eurasian *Homo sapiens* are not the only ones to have interbred with other hominins. Indeed, the genome of present-day Africans has the signature of another human species. *Homo naledi* seems to be one of the best candidates. According to recent analysis, this hybridization, which occurred about 150,000 years ago, would have taken place with a species that had separated from our direct ancestors two million years ago, just like *Homo naledi* (Xu et al. 2017). The interbreeding would then have taken place with one of the last surviving *Homo naledi* representatives.

Finally, a new small-sized extinct human has been found east of the Wallace line, *Homo luzonensis* (Détroit et al. 2019). The remains, some teeth and small bone fragments, have been found in the Callao cave, Philippines, with an age of 50,000 years. It has been suggested that both, *Homo luzonensis* and *Homo floresiensis* used stone tools, but we don't know much about their cognitive capacity. Were they capable of navigating or were they swept to the island by a tsunami? Scientists are now looking for more evidence.

---

[12]The tunnels leading to the discovery of *Homo naledi* were so narrow and curvy that he was only unearthed thanks to young female researchers who were very agile and slender.

2.7 Races and Racism

Our discussion on past human biodiversity and the nature of our interspecific encounters supports our claim that hybridisation, assimilation and replacement were quite common among human species. Nowadays, being left alone to represent the genus *Homo*, how do we deal with this human character? And how do such attitudes determine relations among different populations of our own species? This tallies with a concept still rooted in contemporary culture: the existence of different human races, defined for more than three centuries on the basis of a set of physical characteristics and, sometimes, also cognitive and behavioural traits (Berniers 1684).[13] Our DNA speaks instead of a past in which, due to high mobility, we have always hybridized with different groups, not disdaining other human species.

Today, being the result of all these planetary encounters, our genetic differences are minimal and might be greater between individuals who live in proximity than between individuals who live far away from each other. This does not mean that if we have been living in different environments for generations, there have been no selective pressures favouring characteristics central to survival. This extends beyond the colour of our skin, which has thousands of shades.[14] Other characteristics include height, facial features and body shape. We are all different from each other, but we can't tell how much from our appearance.

Nevertheless, the concept of race is still widespread. It applies, among other things, to forensic science and medicine. We cannot exclude the possibility that someone wants to artificially select us for some unfathomable reason. That would involve eugenics, an alarming field, especially given the power of biotechnology.[15] In this case, the human races would exist but would be an artificial product analogous to the Angus breed for the *Bovis taurus* species, or the dachshund breed for *Canis lupus familiaris*. Yet, even if races do not exist in nature, racism does, and it might be hardwired. For example, the degree of empathy for another's pain depends on their skin colour. Empathy is greater for those considered to be of the same race as the observer but is more moderate for those judged to be of another race (Forgiarini et al. 2011).

In another experiment, however, it was shown that racist biases were modifiable. Using brain imaging, it was found that when people came into contact with ethnic groups other than their own, initially a part of the brain that specializes in negative emotions (the amygdala) was activated. But if the subjects were educated to respect individuals different from themselves, after a thousandth of a second a conflict signal was activated through the dorsal accumbens nucleus, which in turn is activated by

---

[13] Advancing on the previous subdivision suggested by the Bible, which distinguished only between the descendants of Shem, Ham and Japheth.

[14] Nor is this a recent situation. It seems that the variety of our chromatic shades can be traced back genetically to our African ancestors of a million years ago (Crawford et al. 2017).

[15] Think of the CRISPR method for editing our genome. It is proposed to improve the health and performance of our descendants, making them live longer and become more intelligent.

the dorsal-lateral prefrontal cortex. This can dominate negative subconscious reactions. So even if our brain is instinctively racist, a mental process that damps this initial knee-jerk reaction may kick in, as result of a cultural conditioning (Kubota et al. 2012). This is what psychologists call "reappraisal" procedure. No such reaction is foreseeable, of course, if one is taught to fear people different from his own.

A few considerations can be derived from this discussion of our past biodiversity. Firstly, genetic studies and archaeological findings are reducing the cognitive and behavioural differences among human species that we held true just a few decades ago. Secondly, our inter-specific relations have been both friendly and hostile; yet Sapiens had outperformed all other contemporary human species. Why? Finally, there are grounds to check whether the formation of well organised and larger groups was a robust explanation for the extraordinary success of Sapiens in its struggle for survival. We shall follow these lines of reasoning in the next chapters. Then, we shall proceed by discussing the biological and cultural stratagems by which we can form large and complex societies.

Chapter 3
The March for Hegemony

What were the evolutionary solutions of the genus *Homo* in its various forms? What combinations of circumstances eventually led Sapiens to become the most invasive species of the planet? To answer these questions, we must consider the new options that opened up at the dawn of humanity and concentrate on the most important steps that made humans peculiar, with respect to all other species. This evolutionary passage allowed our ancestors to improve their chances of survival without being subject, initially, to a significant modification of their anatomy.

3.1 Tools, Fire and the Environment

At one point the first humans realized that they could equip themselves with stone objects, appropriately shaped with the use of other stones. These implements could be used to scrape the last meat from the carcasses of animals killed by others or to break bones to get their marrow. They could also be used as weapons to kill small animals and possibly other hominins. The first tools were often one- or two-sided flint pebbles with only one sharp edge, formed by knapping one side of the stone with a few strokes. It was thought that the oldest tools of this type, called Oldowan, dated back 2.6 million years. Similar lithic artefacts were recently found in Kenya, near Lake Turkana. They have been dated to 3.3 million years ago (Harmand et al. 2015). This amazing date precedes the first *Homo* remains so far discovered by 500,000 years (Villmoare et al. 2015). The first instruments with sharp edges on both sides appeared 1.8 million years ago and are called Acheulean. These objects were fashioned by repeatedly working the stone on both sides to flatten it. The result was a very sharp almond-shaped blade. In some cases, this involved a long and difficult task that only a few of us would be able to perform today without some practice. But the product was not mere cutlery to get nourishment from the leftovers of others' meals. It was, perhaps, weaponry. Wooden handles made them excellent axes.

The Acheulean instruments are attributed to *Homo ergaster*, the first fully human species, which appeared in Africa about two million years ago. It had a brain of about one kilogram and could run fast over long distances, thanks to its fully arched feet, new Achilles tendons and a stronger femur (Bramble and Lieberman 2004). Natural selection also perfected its hands, building on advances that had made previous hominins better at grabbing and, possibly, throwing objects.

Homo ergaster's hands had an elongated thumb, four shorter fingers, a wide grip surface and all the necessary tendons and muscles (Almécija et al. 2015). They had become precise instruments, suitable for working stone tools and eventually even affixing handles. A stronger, more flexible wrist allowed *Homo ergaster* to grip its versatile tools firmly. It could dig in search of tubers and roots, extract the marrow from bones and strike animals at a distance. The latter capacity was strengthened by advances in the shoulder which allowed the storage of the elastic energy necessary for launching projectiles (Roach 2013). To put it simply, a feed-back mechanism—from tools to body features—was under way via natural selection. More circular self-enforcing, bio-cultural, effects would soon arise.

In the past three million years, strong variations in climatic conditions took place. Colder phases alternated with warmer ones in accordance with cyclical variations in astronomical factors (Raymo and Huybers 2008). During this period, a formidable ally—fire—came into play. Natural fires were very common where hominins lived, particularly in the dry African savannah. Humans soon learned how to illuminate their nights, keep warm, cook food and comfortably sit around it—an exciting occasion to socialise and a habit that humans love to practice even today. Our predecessors also used fire as a weapon. They were not yet competing with large carnivores, but at least they were able to scare them away and possibly steal their prey. They were enriching a diet based on plants and insects, typical in a forest, with otherwise indigestible food, such as meat and tubers. The meat powered the increase in brain size and fire helped combat the parasites and germs in raw meat.

The control of fire could thus be interpreted as a catalyser for setting up and reinforcing a virtuous circle between tools, body anatomy, diet, brain growth and a wider social capacity. The time was ripe to team up and turn from being prey to predators.

All animals feared fire, except for the first humans, who had learned to control it. We do not know how this happened. But we are sure that in order to do it, they had to be bold. They had to imagine themselves "masters of fire" and think of fire as an entity to be tamed. Perhaps the first domestication of a force of nature had then begun. Much later on, Sapiens had to imagine themselves "masters of matter" when taming the atoms and making a nuclear bomb. To achieve this, a lot of brain power (and imagination) had to be accumulated through human evolution. Strangely enough, it seems that this occurred not only by being awake. It was also by being asleep.

While *Australopithecus* and *Ardipithecus* had been sleeping, probably, in shelters built between the branches of the trees, the first humans were already sleeping on the ground. This new habit was very risky, but with new weapons, fire, and larger social groups, one could sleep more comfortably and improve the quality of sleep. This

3.1 Tools, Fire and the Environment

would help to select important brain traits to enhance cognitive abilities. The increase in the rapid eye movement (REM) sleep phases, for example, improves memory and planning (Coolidge et al. 2015). There was the scope for better communication skills when assisted by adequate brain capacities.

When was fire first used and by which species? The oldest human sites, in Kenya, have stone tools altered by heat dating back to 1.5 million years ago (Hlubik et al. 2016). Sites in South Africa with burnt bones date back more than a million years (Gowlett 2016). They probably belonged to *Homo ergaster*, the first human species to leave Africa, about two million years ago, soon reaching Western Asia, China and Southeast Asia. Fossil remains of *Homo ergaster's* Asian variant, *Homo erectus*, date back 1.8 million years. In Zhoukudian, China, their hearths have been dated to about 800,000 years ago. *Homo erectus* could probably aggregate into larger groups of about 100 individuals, doubling the size of social groups of early hominins, which was similar to that of current chimpanzees (Dunbar 2014).

The use of fire was already an advanced practice in *Homo heidelbergensis*, a human with a brain larger than *Homo ergaster's*. The remains are dated between 600,000 and 350,000 years, a period that spans several glacial and interglacial cycles. In France, the hearths attributed to this hominin present burnt bones, blackened stones and even a kind of chimney. The lifestyle of Sapiens was also based on an increasingly intensive use of fire. Gathered around a bonfire, modern humans could make music and probably sing and dance, as well as create beautiful images of Pleistocene animals on cave walls.

However, it seems that the Sapiens could do something more with fire than other humans. Of course, they could alter the environment, like every other species. But modern humans did it deliberately and pervasively. Soon, the domestication of the environment, which started with the control of fire, extended to nature at large. They began with an extensive and selective use of fire to hunt. This practice has not been confirmed in any other human species. The environmental impact of Sapiens is documented in the studies of Lynch's Crater, an ancient lake of volcanic origin in northern Australia. Its sediments hold the secrets of the past 200,000 years of environmental history. Pollen documents vegetation changes. At 50,000 years ago, soon after our arrival, there is a marked transition from coniferous forests to eucalypts—so-called fire-loving trees that need fire to reproduce. These 50,000-year-old sediments also show a spike in charcoal particles (Turney et al. 2001). By burning large areas of forest, the newcomers could coordinate their hunting and control the movements of animals.[1] In doing so they had a dramatic ecological impact on plants and animals. Now eucalypts dominate the Australian flora, and bushfires are a constant threat.

---

[1]Today's Martu, Aboriginal people of Western Australia, still set fires in deserted areas to promote plant growth in order to attract animals.

The geological record shows that megafauna had lived undisturbed in Australia for millions of years.[2] Within a few thousand years of their arrival, modern humans wiped out 23 of 24 known species weighing more than 50 kg. They also decimated countless lighter species. The whole structure of the continent's food chain went awry. Some scholars attribute this disaster to climate change, but the latest archaeological and paleo-climatological discoveries confirm that humans were to blame (Roberts et al. 2016).

In particular, the discovery of many eggshell fragments burned in campfires across the continent shows that our species wiped out the giant flightless bird, *Genyornis newtoni*, about 47,000 years ago. The mechanism was not only hunting and habitat destruction. It was also through eating the bird's big eggs, each of which was equivalent to 25 chicken eggs (Miller et al. 2015). The findings were the first Australian megafauna "kill sites" discovered. The sad story of this species teaches other lessons. Like the panda, the koala and many other present-day endangered species, *Genyornis* had a very specialized diet (Miller et al. 1999), a food strategy that can be fatal if the environment changes dramatically.

Megafauna's extinction in the Americas is also a hot topic.[3] An ancestor of the elephant, the *Gomphoterium*, might have been one of the last victims. Its bones were found near arrowheads attributable to the Clovis culture. They were from a Mexican archaeological site dated to about 13,300 years ago (Sanchez et al. 2014). But the responsibility of modern humans extends to more ancient times. Recently, the bones of a 14,500-year-old American mastodon were found in Florida. They were associated with arrowheads attributable to one of the Sapiens groups said to have colonized the Americas before the Clovis hunters (Halligan et al. 2016). In North America, 34 out of a total of 37 genera of large mammals disappeared. But in this case, climate change played a part. In South America, after the arrival of Sapiens, 50 out of 60 genera of megafauna disappeared.

The great moa bird of New Zealand vanished soon after the arrival of the first Māori between 800 and 1000 years ago. The smoking gun—linking the first humans to forest destruction—was recently discovered in New Zealand lake sediments. The layers from this period reveal molecular markers from human faeces associated to those from biomass burning (Argiriadis et al. 2018). It has been evaluated that a funding group of 500 individuals converted 40 per cent of New Zealand's forests to shrub land in one or two centuries (Perry et al. 2012).

The decimation of animals quickly succeeded our arrival on new continents and islands. They had lived there undisturbed for millions of years, weathering many extreme climatic changes. But the animals were naive: they had not learned to fear us. And we did not seem particularly threatening. We will see that our child-like look

---

[2]We refer, for example, to *Thylacoleo carnifex*, the marsupial lion, to *Diprotodon optatum*, a giant wombat-like herbivore with a camel-like snout, to *Genyornis newtoni*, a huge flightless bird weighing a tonne, and to *Megalania prisca*, a six-meter-long lizard. All these animals went extinct soon after the arrival of Sapiens (Roberts et al. 2001).

[3]With the appearance of Sapiens, *Smilodon fatalis*, the sabre-toothed tiger, disappeared, along with the mighty camel *Camelops hesternus* and other large animals of the glacial period.

and our playfulness would prove central to our survival strategies. In the meanwhile, we have turned into the most dangerous predator and the most invasive species of all continents.

3.2 Once a Player, Always a Player

The extinctions of large animals in Africa are taking longer. They had evolved with us and had learnt to be wary. But now things are changing. It is estimated that most of the large animal species will die out this century. That includes our closest relatives, the apes. The only animals of a certain size that survive around us are those we breed to eat, take us places, work for us, entertain us or keep us company. The rest is made of insects, parasites or pests. We tend to eliminate other creatures through hunting or habitat destruction. But few seem to worry about the consequences even though some contrary voices are rising.

Think of the media clamour, a few years ago, when a Minnesota dentist downed Cecil the lion. This man had paid US $55,000 to the Hwange Park, Zimbabwe, to take another trophy home. The event was perfectly legal. Indeed, some African countries use this "sport"—killing for fun, like children—to raise money to protect natural habitat. The most coveted species belong to the so-called Big Five—elephant, lion, leopard, rhino and buffalo—but there are many others. The killing of Cecil was chillingly cruel. Coaxed out of the reserve with a lure, he was first wounded with an arrow to awaken in the party the emotions of Palaeolithic hunters. He was then pursued for 40 h. Only at the end of this torment was he finally taken out with a shotgun. He was skinned and beheaded. Many photographs document the various phases of this hunting tourism. Some were publicized by the dentist, who insisted that he was not a poacher but a benefactor championing the cause of lion conservation (at least, those that are left).

Cecil had been wearing a radio-collar, which was used by researchers at Oxford University to investigate the decline of these felines (Loverdidge et al. 2016). Their research reveals that between 1999 and 2012, 206 lions died in this "protected" park. Human activities are responsible for the death of 88 per cent of male lions and 67 per cent of females. Most of the males went like Cecil. Most of the females were killed by farmers as reprisals for attacks on their cattle. The effectiveness of "controlled sampling" programs for the conservation of endangered species is much debated. According to the Oxford study, "controlled" hunting has reduced the general mortality of lions but only in comparison to the previous "uncontrolled" hunting. The general context remains that of a population decline due to human causes. In particular, the following reasons should be added to the direct hunting and habitat reduction:

- The killing of an alpha-male forces its followers to slaughter its cubs with serious demographic consequences; Cecil had at least six cubs destined to succumb; one of the survivors, a young adult, was shot down by another trophy hunter;

- The increase in the female population caused by a rule, introduced in 2009, forbidding the hunting of females can cause a genetic weakening of the species through a lower male selection;
- The conservation of endangered species could be achieved much more easily if one were to evaluate the impact of every human intervention on the relevant ecosystem as a whole. In this case, however, the rules of "selective killing" should concern only those species that show an excessive increase in population and involve both prey and predators.

Further doubts emerge about the alleged support that trophy hunting bring to the standard of living of the local populations. It is said that they are given the meat and part of the income received by the local authorities from the hunting activities. But it could be argued that the nature of social problems (hunger, poverty) requires far greater resources than those derived from these revenues. And a disturbing fact should be recalled: the very high mortality of the guards in defence of controlled killing by poachers.

If there is a moral to the story on the fate of the last great animals, it is that we are sensitive to their plight. But the ways in which we protect the creatures, such as the "conservation through commercialization" policy, seem inappropriate, hypocritical and even dangerous. Given that the extermination of large animals is deeply rooted in human nature, only a profound cultural revolution on the ethics of hunting endangered species will have any hope of success. We managed to stamp out cannibalism and overt incest. Perhaps we ought to develop the same disgust for horns and severed heads of large animals hanging on walls with the rationale that it is to protect them.[4] We need to take responsibility for their survival, as adults would do, and stop playing the death game, like youngsters. In the last chapters we shall briefly discuss the very few chances of success of such a cultural perspective.

3.3 Disappearance of the Neanderthals

Going back in time, even the Neanderthals succumbed on our arrival in Eurasia, despite episodes of crossbreeding. Recent studies have restricted the period of our coexistence with them. It was thought that we had lived in the same territory for more than 10,000 years. Based on recent analyses of hundreds of specimens from dozens of archaeological sites stretching from Spain to Russia, however, our overlap might have been less than 3000 years (Higham et al. 2014). This is a very short period, considering that the Neanderthals were a very old and well-acclimatized human species. Our replacement of the Neanderthals happened gradually in a mosaic pattern that forced the distancing of their communities from each other. This

---

[4]This cultural revolution is unlikely to succeed because—among other things—each phase involved in hunting—sighting the prey, killing it, slaughtering it, cooking it and eating it—activates powerful neurotransmitters of pleasure against which the conservationist rationale seems very weak.

3.3 Disappearance of the Neanderthals

exacerbated the biological and cultural separation that had already been going on for some time. The last traces of their culture disappear from the archaeological record between 41,000 and 39,000 years ago. But in present-day Western Europe, they might have already been on the way to extinction 50,000 years ago (Dalén et al. 2012).

Shortly before our arrival, they had been reduced to less than 70,000 individuals with a few thousand fertile women divided into small groups distant and isolated from each other. The genetic analysis of the Neanderthal from the Denisova Cave (Altai Mountains) and the Vindja Cave (present-day Croatia) showed that their heterozygosis (a measure of genetic diversity) was one fifth of that of current Africans and a third of that of current Eurasians. This confirmed that they lived in small and isolated populations of less than 3000 (Prüfer et al. 2017). This lower genetic diversity tallies with previous analysis of different Neanderthal genomes, according to which, between 38,000 and 70,000 years ago, there would have been less than 3500 fertile women (Briggs et al. 2009). Under these conditions, it is reasonable to suppose widespread interbreeding between close relatives, as confirmed by the DNA analysis of a Neanderthal woman discovered recently in the Altai Mountains (Prüfer et al. 2014). It seems that modern humans had managed to avoid this practice, as shown by the study of the genome of individuals buried in Sunghir 34,000 years ago. The results suggest that their network of sexual partners was much more extensive than that of the small group to which they belonged. This is the norm in current hunter-gatherer societies (Sikora et al. 2017).

Genetic weakness aside, we cannot rule out our responsibility in the extinction of the Neanderthals. In northern Italy, scientists have identified the group of Sapiens who could be called into question either directly or indirectly. These people would be the producers of lithic technology classified as Protoaurignacian, which appeared in these parts just 41,000 years ago (Benazzi et al. 2015). We do not know how many modern humans had arrived in Eurasia at the time. Yet genetic and archaeological analyses suggest they expanded dramatically. Their numbers increased tenfold in Western Eurasia in the transition period between the advent of Sapiens and Neanderthal disappearance (Mellars and French 2011). Some say that the Neanderthals went extinct because we took the resources they needed. Indeed, the great herbivores of the Eurasian plains formed part of the diet of both species. But this idea is suspect, for our numbers were still quite small.

We probably compensated for our lesser physical strength with cultural advantages, such as the ability to form large bands and the use of throwing weapons. But these are circumstantial elements. Our success and their extinction are unlikely to be attributable to single factors, such as pandemics or environmental disasters. In fact, it is difficult to believe—as recently suggested—that the demise of the Neanderthals was solely caused by an intense flux of ultra-violet radiation. This would have been induced by the strong reduction of the terrestrial magnetic field that occurred 41,000 years ago. According to this hypothesis, we would have been selected at the expense of the Neanderthals thanks to a genetic variant of a protein sensitive to UV rays (Channel and Vigliotti 2019).

This explanation is not very convincing. We are more inclined to support the hypothesis that our success, and their demise, are best seen as the culmination of a spiral of events combining biological, environmental, social and economic mechanisms. In particular, Sapiens could have started with a slightly longer infancy and greater fertility and longevity; these are crucial parameters to accumulate knowledge, increase population density and build up a complex social organism. To the contrary, Neanderthal women might have been subject to a fertility reduction, triggering a demographic crisis and a reduction of cultural complexity (Degioanni et al. 2019).

Finally, two other important elements can be added to the picture of the valiant Sapiens hunters of Central Europe: a novel food chain and a new ally. This story is an illustration of the advantages that can emerge from the interplay between self-domestication and domestication of another species. We are talking about wolves.

3.4 Man's Best Friend

In present-day Moravia, Central Europe, archaeologists are studying a 30,000-year-old site rich in carnivorous and herbivorous mammal bones. Among the carnivore remains are those of cave lions, wolves, bears and foxes. Herbivore remains include those of deer, oxen, horses, woolly rhinos and mammoths. From the isotopic analysis of collagen in the bones, researchers can reconstruct details of the food chain in the region, now populated by Sapiens only (Bocherens et al. 2015).

In addition to showing who ate whom, the site introduces a controversial character—about 40 specimens of a putative new species of wolf-dogs. These animals differed from wolves of the period and from present-day dogs and wolves. They had big teeth, a powerful sense of smell, a high running speed and other traits typical of good hunters. Their diet was based exclusively on deer meat—an apparently very strange fact—while it seems that modern humans living in the area (Gravettians) fed mostly on mammoth. How is this food specialization explained?

We know that the Neanderthals hunted prey at close range. It would have been difficult for them to down the giant woolly mammoths, even if they could have caught them with the help of courage, great teamwork and a lot of luck. Perhaps they were content to hunt small mammoths, as suggested by the remains in the Neanderthal site of Spy in Belgium, dated to 40,000 years ago (Germonprè et al. 2012). In Moravia and other sites in the region, it has been confirmed that Sapiens could efficiently hunt mammoths of all sizes. They used mammoth tusks and skins to build shelters and other useful objects. How could they bring down these enormous animals with such ease? It required high coordination in hunting strategy and a large number of hunters, characteristics favoured by large bands. Throwing weapons also played a part.

The analysis of the diet of the new species of wolf-dogs paints a much more intriguing picture. Wild wolves fed on a variety of prey. In contrast, the specialized diet of wolf-dogs suggests that they were being domesticated. Why? Most probably because Sapiens fed them venison to co-opt them into the mammoth hunts. This might have been preceded by a phase of self-domestication by some wolf groups. A less aggressive look and behaviour could have let them occupy a new ecological niche. The wolf-dogs might have been tasked with finding and isolating the prey. Modern humans could then hit the mammoth from several directions. If a wounded mammoth tried to escape, they could hunt it, thanks to their team of wolf-dogs. Their new friends could then guard the carcasses—which would have provided more meat than they all could eat—from the hyenas and other scavengers. The first "reserved" food surplus was generated. This was destined to change the structure of society, making it more sedentary.[5]

This reconstruction—confirmed in many ways by older sites elsewhere in Eurasia—is still under scrutiny, but provides us with some preliminary results. First, it suggests that modern humans had acquired a new ecological niche—that of super-predators. If this practice of hunting with the wolf-dogs started early enough, in conjunction with the last Neanderthals, it could have been the *coup de grace* for their definitive disappearance (Shipman 2017). Second, it marks a turning point in our relationship with animals, anticipating their domestication by 20,000 years. Moreover, it introduces an incentive to settle in a specific area long before the agricultural revolution, due to the impossibility of transporting the enormous quantity of food available. This condition would have lasted as long as the availability of extra-large and abundant resources were available and then fade away as populations moved elsewhere. Finally, a new human niche emerged in which Sapiens were accumulating more resources than they needed. The way was paved for the formation of hierarchical societies. In order to do so, man too had to be tamed. What biological and behavioural traits can be envisaged to this purpose? Let's enter the core of our story.

3.5 Self-Domestication of *Homo sapiens*

Besides domesticating other species, mounting evidence is there to argue that we also domesticated ourselves. Self-domestication operates an adverse selection to aggressive traits and leads to a change in the physical and functional characteristics of an organism. It has been studied extensively in wolves, bonobos, and foxes. In such cases, the domestication process slows down biological development by selecting those regulatory genes that influence neuroendocrine maturation (Trut 2001; Trut et al. 2009). As a result, the aggressiveness of adults is reduced. We

---

[5]When the food surplus will come from harvesting (with the advent of agriculture), it will be up to the cats to self-domesticate, to follow the rodents that infested our food. They were already accepted around large Sapiens communities about 10,000 years ago. A few millennia later, they were submitted to artificial selection by the Egyptians (Ottoni et al. 2017).

know that wolf cubs, for example, can socialize much more easily with each other and are friendlier to humans than adult wolves.

The selection of lower levels of aggression could explain new behavioural traits typical of domestication such as a greater frequency of sexual relations, liberation from reproductive urgency and the permanence of a playing attitude in adulthood. It could also explain the increase of all those physiological traits mediated by hormones and neurotransmitters of pleasure. And it would express itself in morphological traits (smaller body size, greater roundness of the physical shapes and forms, a flattened face or snout, larger eyes and smaller teeth) which are less frightening, and therefore lower aggressiveness in the observer. They are all typical of younger subjects (neotenic features) and the female gender (feminization).

This selection would take place by modulating the levels of serotonin and other hormones, which are generally higher in the younger brains than in adult brains. When all these traits tend to be fixed in adults, we talk about pedomorphism. The process of domestication can take place either artificially, when we select those animals that best meet our needs, or naturally, when a species obtains survival advantages in decreasing its levels of aggression. In this case, it assumes a tamer appearance and a more playful behaviour (self-domestication).

Charles Darwin (1871) mentioned the possibility that some features of the domestication process observed in certain animals could be extended to humans. He noticed that tamed animals had more varied physical characteristics than those in the natural state, because of the different reproductive conditions to which they had been subjected. In the same way, the "human races" might have resulted from similar selection process, when restricted by specific environmental conditions. In particular, he noted that in the more civilized nations, the members of the various social classes exhibited greater variability in traits than those observed in the more primitive populations, where greater homogeneity reigned. The influence of environmental conditions (which discriminates between the conditions of survival between classes) might have gone hand in hand with the habit of mating with the members of one's class.

Darwin did not go so far as to state that there had been an intentional process of selection in humans to enhance some particular abilities, except for the well-known example of the Prussian grenadiers. In that case, the tallest women and men of some villages had been encouraged to provide the Kaiser with the best representatives of his high functions. However, if we extend this reasoning to the case of our species, there was no evidence of a deliberate control of our reproduction in general. And most importantly, there was no one else to blame, then.

Today, there are a number of scientific attempts to extend to Sapiens a process of self-domestication similar to that experienced by wolves and bonobos (Leach 2003; Bednarik 2014; Hare 2017). We also changed our behaviour, becoming increasingly pro-social, and our anatomy, by reducing our cranial capacity and our teeth and jaw size. According to this perspective, we would have tamed each other by generating ideas and imposing rules conducive to the formation of social hierarchies. The globular shape of our skull, which emerged between 200,000 and 100,000 years ago, would be the emblem of our self-domestication. But this process might have started earlier.

We mentioned the Jered Irhoud skull attributed to *Homo sapiens*, dating back to 300,000 years ago. His facial features make us think that he was already undergoing a process of "feminization," associated with a reduction of the brow ridge, a shrinking of the face and a reorganization of the vocalization system. It is suggested that these morphological changes reflect a reduction in the average androgenic reactivity (that is, lower testosterone levels and reduced density of androgen receptors), which in turn is associated with an evolutionary trajectory that leads to an increase in social tolerance.

Both domestication and self-domestication can be defined as the result of a syndrome capable of modifying the morphology, the hormonal functions and the neural structure of the tamed. These changes would in turn be sparked by mechanisms that act at both the cellular and genetic level. For modern humans, these mechanisms are guided by a set of genes activated[6] by particular selective processes during the adaptation to our ecological niche.

Is it possible to identify the cellular mechanisms responsible for the morphological and physiological changes that accompany domestication? Some claim it is. The different traits that define the domestication syndrome—all inheritable—could be caused by a very small deficit of cells in the neural crest during embryonic development[7] (Wilkins et al. 2014). In migrating through the body, these cells produce the precursors of many other types of cells and tissues. These include those affecting the craniofacial region, the teeth and the glands that produce adrenaline. It remains to be established which specific genes can be traced back to the generation of that deficit. Indeed, by comparing the genome of modern humans with that of domesticated animals about 40 genes are now emerging as possible candidates for producing such an effect (Theofanopoulou et al. 2017). The human self-domestication hypothesis is therefore gaining momentum.

This syndrome might have intensified during the last few millennia through feedback effects generated by an increasingly protective environment in which we have radically changed the way we got food and clothing, our eating habits and mobility. These responses would have acted both on the conformation of our body and on the functioning of our mind. In specializing in different jobs and organizing ourselves into complex societies, we have become increasingly dependent on each other. This would have made us more delicate and tamer, at least in appearance. Is it with this stratagem that we have avoided succumbing, as a species, to the excesses of aggression that we are able to demonstrate, for example, for economic and territorial reasons? If it were true, one would wonder, by extension, how much self-domestication we would need in the future to get along with each other and survive. We suggest we should not place much faith in this possibility.

---

[6]This activation could take place rapidly through epigenetic processes.

[7]This is a class of stem cells that appear at the start of embryonic development on the crest of the neural tube.

It is well known that our aggression can be both reactive and proactive. The former consists in a persistent emotional and violent social behaviour of single individuals. The latter is about carefully planning acts of violence and attacks for hunting and fighting. We share both attitudes with the chimps and probably it is a legacy of our common ancestor. To expand their social organisms, the Sapiens increasingly needed to neutralise reactive aggression. This is carried out, in a community, through punishments spanning from ostracism to capital penalty. The meeker individuals will be selected for. The neutralisation of reactive aggressors is a very delicate process as it requires keen communicating and planning. It implies the use of complex language and alliances. This would weaken the control of the would-be despots and strengthen the power of peers in maintaining social order.

However, if human proactive aggression persists, for context related reasons, more aggressive leaders could emerge and condition the most domesticated individuals. So the process could go into reverse, and a decrease in reactive aggression could be more than compensated by an increase in proactive aggression. This might explain the "goodness paradox" whereby we are—at the same time—the most gentle and the most aggressive species on earth (Wrangham 2019). Eventually, if a vertical power structure remains strong, individuals can be socially well-behaved while bowing to some dominant figure and its entourage. They could thus remain very aggressive against some enemy, especially if frustrated and trained in combat. Breeders of fighting animals know this well when they submit their champions to all sorts of hardship just to increase their ferocity, and then reward them in exchange of loyalty.

If the survival conditions of our species proved to be more difficult in the future due to some event that accentuated our economic and cultural differences, we should be very careful about entrusting our supposed meekness to settle our disputes. Precisely because we are self-domesticated, we could be very dangerous if subjected to severe cultural conditioning and high material deprivation. And if we are sufficiently angry or frustrated, some alpha entity would not have much difficulty in unleashing our ferocity and gain, at the same time, our devotion, via a widespread dissemination of pleasure and rewards.

Bearing this in mind, let's take a final look at the relationship between power and pleasure. Is it possible that someone else would take advantage of our self-domestication and try to domesticate us, just like we did to some predisposed wolves 30,000 years ago? What kind of intelligence could carry out this last step? And what would be the most favourable conditions (economic, political, psychological and social) to facilitate it? We shall return to these topics in the last chapter. In practice, we will ask ourselves: if we look at our deep history, should we worry about how we allow ourselves to be conditioned by those who give us pleasure? After all, intelligent machines know us very well by now. Could we become, one day, a bunch of children, to be taken care of, like puppies, by a system of intelligent machines? What kind of hardship and what set of rewards would facilitate such a result? Is there a chance that, forced by circumstances—from climate change to increasing inequality—we voluntarily embrace our hybridisation with artificial intelligence and live happily ever after?

3.6 Last Act

The destruction of the natural environment that we mentioned above continued at the end of the ice ages, when some Sapiens groups finally dedicated themselves to agriculture and breeding, reducing biodiversity greatly. The hybridisation with their tools and the domestication of their habitat went on. A more comfortable and sedentary way of life impacted on body features, inducing a progressive gracilisation. It also affected mental features, leading to a growing irrelevance of single individuals and a major empowerment of them as members of a social organism. Indeed, the increasing social interdependence, based on the division of labour, led to social stratifications according to skills, roles and resource availability—all themes that will keep us busy in the next chapters. Fire was used to extract copper, iron and other metals from stone to forge more powerful weapons, to be used this time against each other. The last millennia of our history were marked by genocide as we wiped out the weakest populations of our own species, after passing some of their genes down the stream of future generations.

At present, Sapiens' potential for destruction seems to have no limit. Craving energy, in recent centuries we burned coal and other fossil fuels, contributing to the change of the planet's climate. This trend continues despite global agreements aimed at arresting climate change. We managed to survive the natural climatic changes of the Pleistocene. It remains to be seen whether we'll survive the environmental disasters and mass killings that we can now trigger ourselves. This calls for a few speculative remarks.

As for the near future, population growth and climate instability will be the main causes of human conflict this century. According to the latest United Nations evaluations,[8] the global population will reach more than 11 billion in 2100. But growth will not be even across the Earth. Africa will reach 4.5 billion inhabitants, while Asia's population will stand at 4.8 billion. European numbers will reach only 650 million people, with a high average age. It is in the cards that a wave of young Africans will abandon their lands, battered by aridity, floods and conflict, to populate the North of the world, especially northwestern Eurasia. This will be a new great exodus of Sapiens from Africa. Instead of replacing the Neanderthals, this time the newcomers will fill the void left by the native Sapiens with their low birth rate and aging population. This new exodus has already begun. Many will later abandon the Mediterranean, which is vulnerable to extreme desertification.

The situation will be similar in the rest of the world. The populations of Southeast Asia, driven by extreme weather events and sea level rise, will spill over Northeast Asia and Australia. But they will find few habitable areas there. Similar migrations will start in the Caribbean and Central and Southern America. They will head for North America. No wall will stop them. The demographics of North America will be similar to those in Europe, both in terms of numbers and average age. Conflict

---

[8]Available at https://esa.un.org/unpd/wpp/Publications/Files/WPP2017_KeyFindings.pdf (last access December 2017).

between existing residents and newcomers will probably escalate, giving rise to asymmetric wars. In addition, there will be a new protagonist—artificial intelligence. We will consider the role of this new entity in helping us manage the complexity of our social relationships in the final parts of this book. But first we have to follow, by and large, what happened in between.

Chapter 4
The Naked Ape Dresses Up

Besides all of the above, there are other areas in which human evolution has taken a unique path. For starters, we developed seemingly nonsensical body changes. Think of the innovation of walking upright. It makes us more unstable, exposes our vital and reproductive organs to vicious attack, restricts the birth canal and causes us many ailments as we age. We are then forced to look to advantages that can compensate them: for example, the possibility of running faster, saving energy, seeing further and freeing the upper limbs.

Another intriguing mystery is why we gradually dispensed with our lovely hairy pelts. They had helped us broaden the range of temperatures we could bear, protected us from abrasions we got throughout our adventurous lives, shielded us from the sun and allowed us to wander with our little ones clinging to our mantle, all without losing the use of our upper limbs. What countervailing benefits can be invoked to justify the presence of hair so thin as to make us appear naked? The only mammals that have lost their hair are those that populate the seas, such as whales and dolphins, which apparently don't need it. The fact that we did need it is driven home by our need to wear the fur of other animals in cold climates. In warmer climates, we sport some piece of clothing, too. So, when and why did we lose our hair, and when and why did we start dressing up again?

Indeed these questions are less futile then they appear at first glance. And they are difficult questions because hair is not kept in the geological register, and neither are clothes. To be precise, hair is not totally lost. We have kept it on the head, above the eyebrows, under the arms and on the pubis. It is also on the faces of men. Today, we attach extraordinary importance to our hair for it contributes to our appearance, and baldness often brings despair. But we are ashamed of hidden hair and ambivalent about beards. On the other hand, our body is still covered with a fine layer of hair, very short and thin. It corresponds to five million follicles, about the same as in chimpanzees and many other primates. So, we are not really naked. But since we do not see this hair, or, if we do, we get rid of it, we can assume that it does not exist.

In a bold move, some have recently suggested that our hair loss, due to a single genetic mutation, could be called into play for the development of an upright position (Sutou 2012). According to this view, it would have been difficult for mothers to transport their children throughout the day without having a hairy mantle to which the youngsters could cling. The need to use their arms to carry them around would have freed their upper limbs from locomotion, opening a pioneering path for all of our species, for the use of the first tools. Yet, this would have put mothers at a disadvantage when it came to survival, setting a destiny of dependence of the females on the males and spawning the first family units.

This hypothesis is controversial. To test it, it should be proven that the upright position emerged after the loss of hair. And secondly, it is not convincing when it gives motherhood primacy in explaining female submission to males. If mere survival was paramount, it would have been much more efficient to leave a child in the care of other women—on the basis of reciprocity—than to give a dominant male responsibility for the survival of both mother and child in exchange for submission. We will see that there is a way out of this challenge when women are in a position to select less aggressive males for mating purposes. In the meanwhile, let's look for other explanations.

4.1 The Naked Ape

Let's establish the facts. Among all extant apes—gibbons, orangutans, gorillas, chimpanzees and bonobos—we are the only naked ones. Considering our most distant ancestors, the loss of hair probably did not concern *Ardipithecus* or the other bipedal apes, which from seven to four million years ago still lived in forests. Initially it was thought that the loss of hair began with the last australopiths when they abandoned a predominantly arboreal life and began the adventure of bipedal ambulation in the undergrowth of forests or on the edge of the savannahs. One explanation (Wheeler 1984) was that these traits evolved to reduce heat overload when these hominins were foraging in more open tropical habitats where they were exposed to the direct effects of sunlight.

This viewpoint was challenged by the consideration that it ignored the endogenous costs of heat generated by locomotion (Ruxton and Wilkinson 2011). It was argued that only hair loss could have been selected for the reduction in heat load, not bipedalism. The latter trait had very little thermal advantage and must have evolved for other reasons. These considerations were recently complemented by considering the altitude (around 1000 m above sea level) at which the australopiths were actually living, around the clock, which included the cold nights on the African plateau. It seemed that, at those heights, their pelt remained very useful, evolutionarily speaking. Fur loss could not have appeared, all things considered, until much later. This is probably when the *Homo* populations were living at sea level in open habitats (Dávid-Barrett and Dunbar 2016).

The fact that hair loss occurred along with the multiplication of our sweat glands seems to confirm the need to develop a more efficient cooling system to suit a more arid climate. Under such conditions, water resources would be particularly sought after. A life spent in and around an aquatic environment would have made hair superfluous, when associated with a layer of fat capable of making the body partly impermeable. And this happened as well. One fringe school of thought attributes some of our physical characteristics and behaviour to an "aquatic phase" in our evolution.[1] This controversial theory is supported by a wide variety of arguments, such as our innate ability to swim in the early stages of life, our ability to control our breathing, glandular systems capable of producing fat, tears and the possibility of mating frontally. These characteristics resemble those in aquatic mammals. Such ideas seem compatible with survival in isolation, in a marshy lake or marine environment, such as that prevailing around the Afar Depression, East Africa, about six million years ago. It was only later—it is argued—that we adapted to the drier conditions of the savannah.

According to the supporters of the aquatic theory, the above explanations are consistent with the environmental changes in the heyday of the australopiths. For millions of years, they saw the intermittent advance of the savannahs at the expense of the forests. They were forced to gravitate to increasingly precious water sources. Moreover, sexual selection favouring hair loss might have come into play, as already suggested by Charles Darwin (1871). Indeed, hair loss would make sexual encounters accompanied by caresses more intense, initiating the association of sex with deeper emotions.

Another (weak) explanation can be derived by the presence of body parasites. Less hair favours their reduction, important in that their presence can graduate from a nuisance to a torment, especially in warmer climates. It is therefore ironic that one of our parasites, the louse, could help explain the evolution of our hairy mantle and the introduction of clothes (Kittler et al. 2003; Toups et al. 2011). Lice are very fussy about where they live. Every monkey today has only one species but humans can have three – one for the hair, one for the private parts and one for clothes. This fact bears on the question of when we lost hair.

The DNA of these parasites shows that the common ancestor of chimps' and hominins' lice dates back to six or seven million years ago, synchronized with the split of our lineage from that of the chimp. Since then, hominins and chimpanzees have co-evolved with their lice. However, the hominins caught another species of pubic lice, which had evolved from those infesting the gorillas three to four million years ago. To allow the evolution of two different species of the pests, it is argued, the hominin's hairy habitats on the head and on the pubis had to be separated. This made most think that Lucy was perhaps already "naked," at least on the torso. This idea is now at odds with the studies of the real high-altitude habitat of australopiths mentioned above.

Regardless of which evolutionary process promoted our loss of hair, humans became the first naked primate. Despite all odds, this eventually turned into an advantageous trait for our mobile life in the savanna. But those first naked humans

---

[1] See the symposium "Human Evolution: Past, Present, and Future," London, 2013 (Chiarelli, 2013th; 2013). See also Roberts and Maslin (2016).

had to protect their hairless bodies from insects, solar irradiation and bacteria (Hodgskiss and Wadley 2017; Rifkin et al. 2015). It seems that ochre—a ubiquitous mineral of yellow (goethite) and red (hematite) colour—became the solution to these problems. In Africa, humans could rub it on the skin, as early as 500,000 years ago. And some contemporaries keep doing it today, in some regions of that continent. The use of this mineral went on with Sapiens, who developed efficient techniques to make powder out of it, and use it as a skin pigment. These people were able to make some good use of their two square meters of hairless skin. They could start to represent themselves in colours, and paint messages addressed to those who could read them. An embryo of symbolic thought was developing. Later on, it would be up to clothes to perform this dual function. On the one hand, clothes could protect a naked body from the harsh conditions of the environment; on the other hand, they could deliver a symbolic representation of oneself and the society to which each individual belonged.

4.2 Coats, Shoes and Shelter

Modern humans could have taken their clothes as well as their pests when they travelled out of Africa. Thanks to the availability of new animal furs, such as those of the bear, *Ursus spelaeus*, already a Neanderthal fashion item, it became easier to live in the cold steppes of Eurasia. The cave bear would be extinguished before the end of the last glacial period. Not incidentally, populations of female cave bears began declining, in Europe, precisely around 40,000 years ago, together with the spread of Sapiens in the region (Gretzinger et al. 2019). Apart from being vulnerable to hunting for their fur, they hibernated in caves in which we, the new tenants, wanted to take refuge. Bears eviction could be dangerous but it was worth it.

The Eurasian steppe and tundra were very different from the African savannah. Some warm shoes came in handy. It was found that the Eurasian Sapiens of 40,000–30,000 years ago already wore them (Trinkaus and Shang 2008). This can be shown by comparing human foot bones from this period with those from earlier fossils. It seems that the intermediate phalanges become more fragile with the regular use of shoes. The use of footwear developed new pathologies (Zipfel and Berger 2007). It is curious to note that the bones of the Neanderthals do not have this characteristic, suggesting that they walked barefooted. However, some researchers suppose that they wore shoes made of birch bark (Condemi and Savatier 2016).

With or without shoes, the Neanderthals had developed fairly sophisticated techniques to make clothes of waterproof animal skin. Sapiens, on the other hand, produced pins, needles and other haberdashery to make increasingly elaborate garments. The first use of bone needles emerged in the Eurasian steppes between 40,000 and 50,000 years ago. These tools had probably been invented independently in Siberia and China (d'Errico et al. 2018). Needles evolved in shape and size during the last ice age, and were probably used for sewing cloths of increasing complexity. Dressing was not only useful to protect the body against cold temperatures. We shall see that—together with a number of other stratagems—it would help promote the mechanisms of self-domestication and symbolic thinking.

At some point there might have been a housing crisis. In the absence of caves, and with a great desire to explore, humans needed shelter, especially on the cold steppes of Eurasia. So, they started using the remains of the great animals of the time to build demountable huts furnished with fireplaces (Shipman 2015). Neanderthals are credited with the most ancient housing solution of this type (Demay et al. 2012). But there is only a single isolated shelter that can be attributed to them so far. Traces of many contiguous shelters, reflecting a fairly complex social group, are confirmed only in the case of Sapiens (Iakovleva et al. 2012). Protection from the cold in all possible ways would prove essential to the exploration of Eurasia and, eventually, the crossing of icy Beringia en route to the Americas from the north.

When did we start dressing? And, if we did it in Africa, why did we introduce such a strange innovation, which has never occurred to any other species? As usual, to answer these questions, we must rely on evidence, formulate hypotheses and test them. But it's difficult. If we claim to have direct evidence, even if we can find old-fashioned clothes, at most we can go back a few millennia, as in the case of Ötzi the Ice Man, the fascinating mummified figure we can admire in the homonymous museum of Bolzano, Italy. He was killed while crossing the Alps 5200 years ago. His icy grave delivered up his extraordinarily well-preserved body, his clothes (including shoes and hat) and his weapons and tools. His corpse and these items allow us to partially reconstruct his ancestry and the circumstances of his death. In the tradition of mummy mystery stories, his make us shiver, with tales of strange accidents and sudden death besetting many who came in contact with his body after it was exhumed.

Reaching back much further in time, we will try to establish a date for the advent of a unique costume. How can we satisfy our curiosity? By looking at the aforementioned third species of lice that lives in our clothes (where it lays eggs). From there it moves on the skin several times a day to feed. After giving us a lot of hassles, it can make itself useful and give us a hand in shedding light on this question.

Genetic analysis based on mutations in mitochondrial DNA reveals that the louse that lurks in our clothes separated from the louse we host on the head between 83,000 and 170,000 years ago. This must be when Sapiens began to dress. It is a big spread in dates but it can mean only two things. The first is that our dressing began in Africa before we left. This is hard to believe, given that the climatic conditions then in Africa and in the Middle East were not so harsh. Alternatively, human clothing began in Eurasia, perhaps with the Neanderthals or Denisovans, who had more reason to protect themselves from the weather. In this case, this third species of lice would have evolved on their clothes and infested ours in the close encounters we had with them.

These ideas are yet to be verified. It is puzzling how humans survived the cold of Eurasia before 170,000 years ago. These questions remain shrouded in mystery. But are we really sure that we dressed only because we were cold?

4.3 Dress, Shame and Symbols

Today, we justify our clothing not only as a barrier against the weather but mostly as a social attitude linked to a sense of shame. At a certain point, we are told, we realized we were naked, and immediately we felt ashamed—of exposing our private parts. But these are private only if we dress! There must be another reason.

On the other hand, we cover many other parts of our bodies besides our most intimate areas, which have sexual appeal even though most perform other physiological functions. In recent times, women began to cover their breasts, while everyone covered their buttocks. According to this view, we dress to damp sexual desire, which, in a species that has lost the association between sex and reproduction, could be distracting from daily activities. This thesis seems quite convincing if it focuses on intentions. By covering the parts of the body with sex appeal, we reduce the frequency of sexual urges only temporarily. Indeed, in the long run, modesty could have the opposite effect.

In some parts of the world today, there is no limit on how much a woman should be covered to suppress sexual desire, assumed to be uncontrollable. And it is no coincidence that this desire is bound to be very high when some men are allowed to practice polygamy, leaving other men unable to afford a wife. In some polygamous societies, as well as covering her trunk, a woman must cover her legs, arms, hair and face. Even her eyes must be hidden behind a grid of fabric. The woman becomes an imagined and submissive reality. She can thus turn into a slave, of which man can dispose of at will, having bought her services from her family. The husband can even send her back if she does not perform or disobeys.

In contrast, in so-called "modern western" culture, many women expose most of their bodies. Although this might enhance their sexual appeal, the signal attenuates. Consider the milieu of nudist colonies. After the initial shock of seeing naked bodies warts and all (bodies that cannot be idealized), one gradually becomes inured to them. And people in the primary societies of today are indifferent to nakedness.

So, setting moral norms apart for the moment, it seems that the intensity of our reactions against nudity—when it implies desire without realisation—is conditioned by at least two elements. On the one hand it depends on the degree of sexual satisfaction that men can achieve in everyday life: the higher the satisfaction, the lower the stimulus. A high presence of involuntary bachelors is a good proxy for sexual stress, and therefore a higher sensitivity to female body exposure. The first two main causes for such a case are: (1) cultural selection against female babies before and soon after their birth, in societies that prefer sons over daughters; and (2) the presence of polygamy, which enforces celibacy to the less fortunate. On the other hand, sensitivity to nakedness depends on the frequency of exposure of the naked body to the observer: the higher the number of exposures that men have to "endure," the lower the sexual stimulus of subsequent exposures.

Entering moral norms, the view that women's clothes are designed to suppress the sexual stimulation of men bears institutional consequence. It amounts to the transfer of responsibility in negotiating sexual transactions from men to women. If a woman

4.3 Dress, Shame and Symbols

does not comply with some rules of modesty in her dress and behaviour, she will be blamed for rape or harassment, and the man will be excused. It is often said that rape is more about power than sex. Now we say that compulsory dressing is about a transfer of responsibility from the perpetrators to the victims, in the absence of sexual consent. Power here is associated with impunity. Dressing assumes an institutional function: it is a rule to be observed and enforced in a male-dominated culture. Along with other restraints, it becomes an instrument for the domestication of women. The situation can be compared to keeping a dog on a leash, but it can also extend to genital mutilation and death.

Later, we will pose another question: why, at some point, have we dressed independently of the environmental imperative to do so? The answer could rest on the image we want to project. Its origin can be explained by an expansion of the mechanisms linked to sexual selection (competition for mates). People reach out and try to seduce, in a broad sense, by drawing admiration and respect and invoking identification. By dressing, we relate socially and can communicate whom we wish to be perceived as. Ultimately, if this fictional character becomes real, in the mind of the beholder, we can generate roles, hierarchies and relationships, and produce meanings that we have created. More than dressing, we are wearing masks. We create a character that, thanks to the clothes we wear, can be identified as belonging to a youth tribe, a social category, a religious faith or profession. Defining oneself with an *ad hoc* image will be one of the many ways to connect socially and produce important effects, both tangible and intangible.

It is to this ape, once naked and now dressed in clothes and culture that we will turn: an ape who, after getting rid of its fur, thinks that it could also cut loose from its "animal spirits." We will focus on this theatrical character and on some of his stage costumes in the next chapters. The plot will develop in an increasing dependence of modern humans on things and other people. The result will be a choral performance of Sapiens as members of the societies they can imagine and promote.

Chapter 5
The Evolution of Woman

We have always represented human species in the masculine case when discussing the evolution of our genus. Though referring to both genders, the term *Homo* seems particularly unsuited to the women of the genus. An attempt was made to correct this imbalance. Referring to the aquatic hypothesis, Elaine Morgan (1972) proposed a narrative of human evolution focused on the role of women. But her contribution was immediately consigned to the dustbin of anthropology.

The history of our origins refers mostly to males in the context of the evolution of the brain and dexterity. Women remain in the background. If anyone points this out, it is claimed that references to the man include the woman (but never vice versa). In addition, we speak of the "thought of man" and the "aspirations of man" with the implicit assumption that this means both genders. At least this is how the story goes. In the term, "hunters and gatherers," it is assumed to be up to the hunters (whom we imagine to be men) to provide the main course. The gatherers (whom we imagine to be women) seem to provide only the side dish and dessert. In sexual competition, contrary to many other species, it is believed that women have to compete to attract a male and keep him close through their beauty, grace and meekness. As if they had to overcome a handicap. In the rest of the natural world, however, it is almost always the males who have to win the favours of the females. They are the ones who adorn themselves in the most beautiful feathers, dance complex dances, sing, chase and fight. Females sometimes run away or pretend to do so, but generally wait serenely for events to unfold.

We also think that it was the male who perfected "his" stone tools, making them more and more deadly. But women could also hunt prey. And they could use the tools to catch small animals, fish, search in burrows, rummage in the ground, and collect and chop plants and roots. They could provide all kinds of delicacies. It would make no sense to think that the males disdained these little chores. And if they did, who would have invented and perfected these tools? Always the males? Based on what experience?

If we include women when we speak of men but not men when we speak of women, we might overlook important aspects of our evolutionary history. So, let's focus on women for once, using this term for all hominin females, which are well represented in the fossil record. Indeed, the most famous findings are often of females, not males. For example, Ardi, Lucy and one of the *Australopithecus sediba* (Berger 2010), as well as the first Hobbit to emerge in Liang Bua, were all female. Also, Denisova's *Homo* is a woman, as well as many finds of *Homo erectus*, *Homo neanderthalensis* and *Homo sapiens*.

How does one recognise gender in a few fossil bones? Before DNA analysis was extended to more and more ancient fossils, we have long been limited to a morphological analysis, which remains valid. There is sexual dimorphism, especially in the pelvis. In adult hominins, especially in the most archaic, the female was much smaller than the male. In *Australopithecus*, for example, the males reached almost 170 cm tall while the females remained below 130 cm. Dimorphism is much less important in modern humans. Finally, female bones are generally thinner and lighter.

By focusing on women, we shall see that important details emerge about Sapiens' evolutionary path. Indeed, it was thanks to women that our species could turn some difficulties into an extraordinary series of cumulative advantages.

5.1 Giving Birth: A Risky Business

A singular characteristic in human evolution is the progressive increase in the risk of death at birth, both for the unborn child and mother. The adaptation of the pelvis to bipedal locomotion depended on the muscle reorganization necessary to keep the body upright. The new structure of the pelvis, though facilitating bipedal locomotion, increases the risks of procreation, narrowing the birth canal. In bipedal species in which the brain tends to expand, mothers are hit with a double whammy. Childbirth takes on a complicated dynamic.

For Ms. *Homo ergaster*, giving birth was already a difficult matter, compared with that of the *Australopithecus* (Simpson et al. 2008). For Sapiens, it became even more difficult. The foetus has to make two half turns to get out of the womb, a complicated manoeuvre. The mother needs help to deliver her child alive and, possibly, survive herself. The advantages of the upright position and the increase of the cerebral volume came at the price of greater female and child mortality. Evolutionarily speaking, it was a recipe for disaster.

In different human species, this obstetric dilemma found a very effective evolutionary solution (Rosenberg and Trevathan 1995). The risks associated with reproduction could be mitigated by anticipating the time of birth. The complete development of the baby's brain (and hence of his head) could take place partly inside and partly out the womb.[1] A value has been calculated for the maximum size

---

[1]Perhaps the shortening of pregnancy had already begun with the pre-human hominins. From a study of the 3.3 million-year-old remains of an *Australopithecus afarensis*, discovered at Dikika in Ethiopia, we can see that she was already less developed than chimpanzees. At about three years

of the baby's brain to enable it to be born alive without killing the mother. It is 500 mL. Extant newborn Sapiens tend to stay below this limit with an average volume of 350 mL. Lower volumes compromise the child's survival.

The natural selection that favours women able to give birth prematurely (but not too much) has a minimal cost: the need to care for premature babies. An acceptable sacrifice, and perhaps not even a sacrifice, when compared with risking one's own life and that of one's child. Those who pay the highest price are women who die in childbirth or from complications of pregnancy.

According to some scientists, the adaptation of the female pelvis to the neonate size has been less effective during the last 50 years. During this period, there has been a high rate of caesarean sections. The procedure boomed in more developed societies, where the unusually high rate casts rising doubts about the need for it in all cases. This has been related to the increased frequency of obstetric complications. According to a study at the University of Vienna (Mitteroecker et al. 2016), this new course relates mainly to a mismatch between the size of the foetus at birth and that of the mother's pelvis. The scientists say foetus-pelvic disproportion is the result of adverse "obstetric selection" compared with natural selection.

Since Caesarean section takes place regularly on a large scale, a 10–20% increase in foetus-pelvic disproportion rate can be predicted by a mathematical model. This result depends on the fact that mother/child mortality is reduced in all cases that exceed a critical threshold in the distribution of the probability of survival. Thanks to the diffusion of caesarean sections, the tendency to give birth to bigger and more robust babies by mothers with a narrower pelvis is increasing. These traits will pass down the generations. In the future, birth assistance could become essential.

Furthermore, a recent study shows that there is a substantial variation in the present-day shape and relative size of the birth canal of women from different regions of the world; this is a consequence of genetic drift and climatic adaptation. Accordingly, human pelvic adaptation to evolutionary pressures deriving from bipedal locomotion and human brain growth ought to be rethought (Betti and Manica 2018).

A final oddity, typical of human females, is the presence of menopause. This is a weird evolutionary trait seemingly at odds with Darwinian survival of the fittest. There are at least two possible explanations out there. Given the extreme dependence of newborns on their mothers and the dangers of pregnancy in later years, menopause could be considered an evolutionary advantage. In other words, if the survival of the youngest critically depends on the survival of the mother, the mother's reproductive capacity must be shorter than her life expectancy. This solution centres on the care needed by offspring. On the other hand, a woman who ceases to be fertile can help fertile women with their offspring. Menopause would thus benefit the whole species. The two arguments have been compared but are still to be confirmed

---

old, her brain measured only 330 mL, corresponding to about 70% of an adult brain. At the same age, chimp brains measure 90% of that of adults (Alemseged et al. 2006).

(Shanley et al. 2007). However, we don't think that, for the emergence of menopause, the previous explanations are mutually exclusive.

What considerations can be drawn from this short review of the problems our species had to face during its reproduction activities? It seems that females have been struggling along a very narrow evolutionary path, and that circumstances have arisen to favour a longer infancy and pro-social behaviours (within and across genders). These features would help develop a number of positive and cumulative effects, starting from the transmission of knowledge and the development of cultures.

Focusing on the Sapiens newborns, the stratagem of modulating the maturation times extends from the postnatal period, through childhood, to adolescence. Today, we are the only primates with such a long relationship between parents and children. This feature confers a powerful evolutionary advantage. A brain growing rapidly outside the womb can absorb an enormous amount of information. It is very receptive from a social and an environmental point of view. It is malleable and easily conditioned when one wants to impose ideas and modes of behaviour on it. What is learnt in childhood remains etched in our minds. And in the first few months of life, with the help of our senses, we can develop a greater knowledge of the world than we could have had we still been in the womb. These advantages were crucial to the accumulation and transmission of culture.

Even the development of language (spoken and sung) can be attributed, at least in part, to the relationship between mother and child. In primates, physical contact is a demand of the little, insecure ones. The mother's murmurs of reassurance free her up to focus on other tasks (Falk 2004). Through natural selection, this ability to communicate at a distance would be transmitted and reinforced as the necessary brain structures evolved. The hypothesis that women invented language is consistent with the observation that, in a society that remains promiscuous, and therefore devoid of the father figure, the responsibility for rearing the offspring falls mostly on mothers. They became the main conduits of knowledge and promoters of cognitive abilities.

5.2 Monogamy or Polygamy?

The expansion of Sapiens populations, followed by an increasing complexity of their social groups, begs the question of their sexual and reproductive behaviour. Without archaeological evidence, there are different opinions, most of which are conditioned by present-day cultures. But to get an idea, we could start by observing the species that are most closely related to us.

According to some, Sapiens, like gibbons, formed monogamous relationships since the time they emerged as a species. Others believe that polygamy was the norm. But what form of polygamy? The patriarchal, typical of hierarchical structures

5.2 Monogamy or Polygamy?

dominated by an alpha-male as in chimpanzees,[2] or the matriarchal (polyandry) seen in the bonobos, in which the group is led by females which allow for varied and intense sexual activity?

We have already touched on our genetic affinities with these latter two species and mentioned the dual nature of our social behaviours. However, still quite in the dark, we know that male polygamy is generally associated with greater sexual dimorphism, typical of the more archaic hominins. That is to say: compared with females, males were much bigger and robust. This feature remained to a lesser extent in Neanderthals and Sapiens. It could be argued that, since we experienced this behaviour in the past, some of its aspects would be preserved, even without much stronger males. Now males can dominate females through other types of conditioning—social, economic and religious.

The thesis of our innate tendency towards monogamy is based, instead, on the idea that, once turned into hunters of large animals, the division of labour between men and women (the first absent for long periods, the second at home to look after the offspring) would have favoured monogamy and affection of an exclusive nature in order to discourage sexual incursions of passing males towards settled females (Morris 1967). This idea is corroborated by the importance our species gives to sexual activities as a bond between individuals. Having disentangled sex from strictly reproductive motivations, we would have encouraged the rise of an extraordinary sexuality compared with many related species.

From a different point of view, humans would have had, at the onset of their social behaviour, various and variable sexual relationships within a group of individuals who knew each other very well and lived together for long periods. In this case, the paternity would have been a collective fact and the children would have been bred within the community. In principle, thanks to advanced DNA analysis, on the fossil remains of our ancestors, it would be possible to shed light on this issue. But we suspect that, even in the past, we were undecided on which model to adopt. We have already stressed how varied and variable our behaviour can be, and we have attributed the source of several evolutionary advantages to the variety of our organizational models. It could have happened the same way in the sexual sphere.

If this were the case, and genes were passed on from all kinds of sexual arrangements, according to cultures, then we would witness a tension between the development of societies that are more permissive and promiscuous and the development of more judgemental and repressive societies. This situation is bound to be a cause of clashes—especially when behaviours are strongly linked either to ideologies or religions—when people belonging to different cultures meet and merge.

---

[2]To be true, the hierarchical order of the chimpanzees is flexible and composite. Females have their own hierarchical orders and males make frequent sexual incursions into the female population (https://sciencing.com/chimpanzee-mating-habits-6703991.html).

5.3 Taming the Female

An enduring notion is that women are more docile than men and have a natural tendency of subordination to them. Indeed, this condition is observed in many parts of today's world, but we don't know for how long and why. One could assume that this phenomenon is part of the broader process of self-domestication that affects our species as a whole.

We have glimpsed these matters in the context of the transformation of wild wolves into wolf-dogs. That domestication was initiated by a phase of self-domestication when some wolves gained an advantage through a softening of their traits and becoming submissive to humans. A bold analogy between wolf-dogs and women—no offence—is tempting, for even women sometimes see the advantages of self-domestication around dominant males. However, this topic is difficult to treat rigorously. It involves the confluence of many opposing forces—psychology, instinct, economic imperatives and power. And it is impossible to rank these interacting spheres in a hierarchy, at least not in terms of their effect on everyone. However, different societies and cultures do assign an order of priority. Let's try to think about some possible intersections with a cool mind.

Consider, for example, how the loss of some degree of freedom on the female's part interacts with the male requirement for certainty of paternity. She has a wide choice of sexual partners and she knows it, so when she grants a male a right of sexual exclusivity, she makes a deal that does not seem at all advantageous for her. She starts from a position of (information) advantage—she has no doubts about her maternity—but in practice she is satisfied with a promise of mutual fidelity (in the best of cases). How can we account for the woman's strange behaviour? From a psychological point of view, her supposed self-domestication (here meant as a reduction in her degrees of freedom compared with "wilder" behaviour) could be achieved through a mix of domination and fascination on the part of the man. This mix can satisfy men, too.

Probably, a powerful catalyst for both genders comes into play, one that relies on the production of several endorphins and is called love. Love is a condition of euphoria, albeit temporary, in its most intense form, but we sometimes choose to consider it as a fundamental pillar of our relationships, thinking that it will one day transform into something more peaceful and lasting. This means faith that a hormonal cocktail in which the exciting effects of the neurotransmitter, dopamine, initially prevail, will work later as well, when the calming effects of oxytocin and serotonin take over. Otherwise, when a taste for dopamine prevails, unions tend to have a limited duration and are replicated in series.

The commitment to exclusivity is subject to opportunistic behaviour (the notorious betrayal). In the case of women, however, her "cheating" generally elicits greater social disapproval, which can range from mere disdain to death by stoning (in largely polygamous cultures). For males, society is much more permissive, with the reaction ranging from envy to indifference (except, obviously, for the woman affected). There is therefore an incentive for women to honour the loyalty bond (for they get more social conditioning) and therefore submit to a new (cultural) asymmetry in which males who break their promise of faithfulness (or not, if polygamy is allowed) do so with impunity.

The economic dimension of self-domestication is much simpler. In a stable couple, the woman promises to provide various services indefinitely and free of charge. This is a huge advantage for a family unit and for society as a whole. It has been calculated that, if paid under today's market conditions, these services could be equivalent to, and even exceed, the average per capita income of a developed country. In return, the woman gets a standard of living according to the husband's capacity. She might contribute to the standard of living of herself and her husband if she does outside work. Another asymmetry is set up if she still shoulders the burden of services to the family. This kind of asymmetric exchange can become a trap, if it turns into the enslavement of the woman over whom the man claims proprietary rights: a trap that can sometimes becomes deadly when the man does not accept abandonment.

On the other hand, the benefits of self-domestication are minor if women are given the opportunity to provide for themselves and their children without having to rely on the support of men. This scenario is much rarer. Female domestication is a strong social and cultural matrix varying over time and space. Today, these behaviour patterns can also apply to men when the woman is the main breadwinner. All of the above is obviously an extreme simplification of reality. There are many women who are not tamed at all. And there are many types of families in addition to those just described. Yet, with these brief considerations, we hope to have shown how difficult it is to reduce any such behaviour to the "intrinsic nature" of the female spirit.

We could conclude that, although a certain degree of woman's domestication might benefit some of them when it is the result of a conscious and utilitarian choice, the subjugation of one gender to the other is strongly determined by socio-economic conditions. It turns out to be an element of weakness when, focusing exclusively on the reproductive functions of women and the supply of unpaid work, it reduces the physical and intellectual resources that they could provide to society. And it is a source of great unhappiness when it reduces non-consenting women to mere sex objects to whom the best drug of all—emotional involvement—is denied. However, being a cultural phenomenon, we could get rid of it if we agreed to do so.

It's worth saying a few words about the supposed and innate cognitive inferiority of women to men. This is the main justification for a certain degree of domestication imposed upon women by a patriarcal society that denies education to women. So far, there is no evidence for this. And indeed, it is hard to find any evolutionary advantage in such a possibility. Some recent studies do confirm that the cognitive abilities of the two genders differ but are by no means superior or inferior according to some arbitrary scale. Men out-do women on some specific cognitive tasks, while women out-do men on others. There is thus scope for cooperation between men and women in complex problem solving. It also seems that the levels of the hormones testosterone, oestrogen, and progesterone affect cognitive function. These vary with the cycles that drive men's and women's daily lives (Upadhayay et al. 2014).

On the other hand, statistical data—i.e. the proportion of women occupied in "higher activities" in the economic and social ladders—are notoriously biased against women, even in most modern societies, and cannot be used as evidence of an assumed cognitive gap. Indeed, uneducated women denied the opportunity to form good social relationships have arrested development, especially if they are poor in the family and work contexts. But the school and work performance of women does not, at the moment, differ from men's, other things being equal. Finally, girls may be educated to adopt a cultural model of submission to the male; but there are also narratives that tend to enhance the creative, instinctual and savage components of women (Pinkola Estés 1992). The archetype of Cinderella coexists with that of Bluebeard.

From a different perspective, there might be some other qualitative differences between men and women. There is a bulk of evidence, in modern societies, to support the idea that gender difference does exist and that males tend to be more dominating, self-assured, aggressive and independent, and that women tend to be more sociable and sensitive to feelings. Often this breaks down to asserting that men are more "rational" and women more "emotional." But it is unclear whether these observed differences are biologically driven or rather the result of social expectations, enhanced by training and rewards. Yet, there do seem to be some biological differences in male and female brains, when we consider that the cerebral cortex is divided into two hemispheres with different functions. The left hemisphere is mostly associated with analytical and sequential tasks, the right hemisphere with intuitive and emotional tasks. However, emotions are not located in any particular area of the brain and are the product of a complex set of interactions that we are still far from understanding thoroughly. It has been suggested (LeDoux 1998) that a "limbic system," located in the forebrain, is the home of emotions and that a particular section, the amygdala, activates when someone is subject to emotional stimuli. This automatic response is common to all species. In humans, to select for the most appropriate reactions to emotions, the prefrontal region of the cortex is also activated. *Feelings* thus emerge as a result of our brain's interpretation of emotions (Damasio 2004).

Drawing on this distinction, which goes often unnoticed in current debates, is there a biological difference between men and women in their predisposition to feelings? This topic is very controversial, not only because it touches upon sensitive preconceptions—think of the notorious idea that men are less sensitive than women to emotions—but also because a number of neuroimage studies provide conflicting results on the morphology of male and female brains (Bishop and Wahlsten 1997).

Some anatomical studies tell us that female brains are more symmetrical than male brains and that the *corpus callosum*, the main bundle of nerves that transfer information between the two hemispheres, is relatively larger in females than in males. This would suggest that women have a higher propensity towards feelings with respect to males. However, as the genesis of feelings is thought to be located in the right hemisphere, and their expression in the left, it may well be that men and women have the same level of feelings but that women are more efficient in expressing them, thanks to better links between the two hemispheres.

In a recent study, deep-learning neural networks have been used to study morphological trends in a sample of 490 men and 575 women, confirming that gender-related differences exist in several brain regions, including parts of the praecuneus, of the frontal lobe, the thalamus and the cerebellum (Xin et al. 2019). However, morphology cannot be the end of the story.

When asked, men do report to be less emotional than women, but again it is unclear to what extent this is the result of what they are educated to think of themselves. So, even with relevant differences in the brain's make-up, we still don't know how much of the difference between emotions and feelings in the two genders is due to biology or rather to cultural drivers. But if a difference is found, whatever the source, one is tempted to conclude that the gender more sensitive to feelings is also more exposed to be subjugated by a manipulation of them. This will bear serious consequences for our future evolutionary trajectory.

5.4 Sexual Selection: The Role of Women

Previous considerations were based on a daring extension of the first (alleged) self-domestication of wolves to women's condition in some modern societies, stressing the relative advantages and risks of a choice that are subject to different degrees of freedom, according to context and cultures. There is one byway, quite neglected, traversing the role of women on our evolutionary pathway. We have emphasized that sexual selection is almost always a male matter in other species. In most cases, males compete to pass on their genes. They use the most imaginative strategies, even to the point of endangering their own lives to conquer as many females as possible and maximize the spread of their genes. Consider the vulnerability of the male peacock to predators.

For humans, it is often claimed that these characteristics extend from the physical (symmetry, strength, agility and fitness) to the mental (ability to communicate, to build increasingly sophisticated instruments, to be generous and empathetic, to be imaginative). They are known in the common lexicon as the "superior capacities" of humans. If sexual competition referred only to males, however, it would give females a key role in evolution. Women could select the main anatomical and behavioural traits of our species. They could also choose our social structure if the definition of "fitness" encompassed the social position of the male. And even be blamed, as in a mirror image, for the construction of hierarchical societies based on wealth and male power. The evidence, however, tempers these notions. In contrast to what happens in many other species, we also see sexual competition among women, as hinted by their efforts to appear more attractive according to strictly male priorities. This is by accentuating some physical characteristics (the shapes and size of some parts of the body) and some mental ones (acquiescence, for example) that seem particularly pleasing to males.

There is thus an ambivalence in our species which keeps being reiterated. We argue that by mixing all the physical and mental ingredients mentioned above and seasoning them with feelings, we generate an important component of our evolutionary path, based on sexual selection in the broadest sense. This combination of masculine and feminine characteristics has made us all more seductive (sexually and socially) during the formation of our social pyramids. So far this has been an element of strength, useful for the formation of increasingly complex societies. But perhaps a glue that is also made of emotions could turn into an element of weakness in the years to come.

One last question is worth considering, namely the role of women in the general process of self-domestication of Sapiens. The behavioural, physiological, genetic, neural and morphological characteristics that are presently referred to as "domestication syndrome" (Leach 2003, 2007) were discussed earlier. Let's now consider the three major mechanisms that anthropologists indicate as potential drivers of human self-domestication. The first one highlights the social benefits that promote higher reproductive fitness for less aggressive individuals (Cieri et al. 2014). The second one stresses the collective advantages that arise when banishing excessively aggressive individuals, thereby preventing their contribution to future generations (Wrangham 2014, 2018, 2019). The last one suggests the prospect of sustained female choice for less-aggressive mates, who are more likely to invest energy and resources toward shared parenting (Cieri et al. 2014). Let's focus on the latter mechanism.

We have touched upon this possibility when associating bipedalism with hair loss, a combination of events that strongly disadvantages women when it comes to raising a child. We have said that the solution of granting durable and exclusive sex rights to a dominant male was probably less efficient than cooperating with other women in child care when foraging. But what if there was another solution? Very recently, the women's selection of less aggressive males has been statistically tested for context-dependent females on the base of sexual dimorphism. Male and female stature data for 28 societies were gathered from the Standard Cross-Cultural Sample, and multivariate regressions were applied to examine the hypothesis. The relative level of self-domestication (indicated by sexual dimorphism) has been shown to vary across human societies in accordance with women's ability to exercise mate choice in their social context. It is thus confirmed that context-dependent female mate choices significantly contribute to lower sexual dimorphism, suggesting that their mate choice has played an influential role in human self-domestication (Gleeson and Kushnik 2018).

Chapter 6
Work, Leisure and Learning

We have seen that biological childhood and life expectancy lengthened throughout human evolution. This is supposed to give humans more time for recreation and leisure. It would also help young people to learn more from the elders. In this chapter, starting from what we know about the different age phases of hominins' life, we will briefly discuss how work, leisure and learning developed in the past, and how all this helped to form ever larger societies.

The distinction between work and leisure time is very recent. It was fully developed during the last industrial revolution when someone called the employer, by providing the tools and machines to work with, became the owner of the goods and services produced by his employees, whose time he bought in exchange for a salary. Before then, artisans (who owned the tools and the machines) often mixed work with leisure, while the work of apprentices was offered mostly to learn a profession or trade. And traders roamed the planet "working" by travelling and cutting deals. Work could also be provided on demand free of charge, in the case of slavery, or to get what was left once a rent was paid to those who claim an exclusive right of exploitation of some natural resource (for example fertile land). Work and leisure time were intertwined.

In a society of hunters and gatherers, such as the one that has characterized much of our deep past, the distinction between work and leisure did not make much sense either, and the distribution of common products was based on availability. Work, however tiring, was often a source of satisfaction, and the sharing of what was hunted and collected was not directly linked to the contribution of each member of a group. Dividing on an equal basis the collective outcome of a day's work is the most efficient way to get along, because this attitude reduces for each individual the risk of remaining with an empty stomach after a bad day of hunting or gathering. This is still the attitude of present-day Australian Aborigenes and other indigenous peoples.

But how was life in general along our evolutionary lineage? Is it true that a great deal of time had to be dedicated to the struggle for survival? And on whose shoulders did it fall, if we consider different age groups? To answer these questions, let's begin by seeing how long the different age phases were for our ancestors, and what kind of behaviour and opportunities they encouraged. Later we will consider some archaeological evidence on how our ancestors spent their "free time".

6.1 Growing Up Too Fast?

The child of Taung, a three-million-year-old *Australopithecus,* already had an erupted molar. According to the Sapiens parameters, he should have been six years old. Counting the growth lines of dental enamel with x-ray microtomography, however, it was discovered that he was only three-and-a-half years old. Pre-human species quickly passed from infancy to maturity, with a reproductive age earlier than ours. Their lifestyle was probably similar to that of present-day apes.

Archaic humans also developed faster than modern humans, as confirmed by Turkana, a young *Homo ergaster* found near Lake Turkana, Kenya, whose skeletal remains have a geological age of about 1.6 million years. At the age of nine, he had a 900-millilitre brain and was already almost 160 cm tall. Despite having the skeletal structure of a present-day thirteen-year-old, the growth lines of his dental enamel suggest he had a biological age of only nine years. He too had a very short childhood. On the other hand, there is no sign that his species was creative in the arts or in the production of new tools, having continued for over a million years to reproduce the same Acheulean artefacts. However, the quality of the stone materials, their manufacture and symmetry, slightly improved over time. It is thought that their increasing "beauty" could be associated with some form of seduction or tribal prestige. In any case, the construction of these objects required an extensive plan of work, which had to be transmitted to the following generations, partly by imitation, and partly by some rudimentary verbal and gestural means (Wynn 1995). And a number of tricks had to be passed on, to control fire and its applications in cooking and hunting. There is no evidence that *Homo ergaster* was endowed with some degree of symbolic thought, even though it is difficult to preserve this kind of proof for such a long time. Its main sources of fun were likely to be mostly limited to food, sex and the rapid rearing of its offspring.

Its Eurasian descendant, *Homo erectus,* was the first, as far as we know, to produce engravings with a geometrical pattern on shells. The evidence is from the island of Java, Indonesia, dating back half a million years (Joordens et al. 2015). According to the discoverers, these signs may have had a symbolic character. If true, that would push the advent of this mental trait, presently attributed exclusively to modern *Homo sapiens,* back in time. In any case, having a much shorter childhood compared with ours, the accumulation of knowledge was still limited.

A long childhood has been confirmed for *Homo sapiens* right at the beginning of its emergence as a species, a very useful trait for the transmission and accumulation of knowledge. This is suggested by the micro-analysis of the teeth of an 8-year-old boy found on the site of Jebel Irhoud in Morocco, which date back to 300,000 years ago. The enamel layers tell us that his biological development corresponded precisely to what is expected today at that age for present-day humans (Smith et al.

2007). Indeed, the adults of these early Sapiens were also starting to maintain juvenile anatomical traits, such as a smaller face and weakness of the brow ridges, compared to previous humans (Hublin 2017). This suggests that self-domestication was already under way.

6.2 Art and Entertainment

At some point it seems that not only our anatomy, but also our mind, maintained child-like traits that extended to adult age. Our high propensity for play became a distinct characteristic of our species. There is large archaeological evidence that our ancestors started to spend increasing time and energy performing collective rituals, which included sharing images and music, for sheer fun and pleasure. Transmitting information about the "real" world for practical purposes was not enough. They needed to amplify it with symbols and representations. They were also "augmenting" their body with ornaments and signs on their skin.

In Africa 80,000 years ago, some Sapiens began to draw abstract signs on ochre tablets (Henshilwood et al. 2002) and to create elaborate shell necklaces. Traces of a similar behaviour are found throughout the eastward expansion of modern humans. In some caves of Sulawesi, in Indonesia, archaeologists discovered rock paintings of handprints and animal figures dating back to 39,900 and 35,400 years ago, respectively (Aubert et al. 2014). In a cave on the adjacent island of Borneo, a figurative painting of an animal, possibly a wild bovid, was found and dated to 40,000 years ago (Aubert et al. 2018). An astonishing display of stencil hands—often very fine and gentle—were scattered all around. To present-day minds, they are like a Palaeolithic selfie, or signature. Banksy, a contemporary street artist, can claim to belong to the most ancient tradition of wall art.

On reaching Europe, other Sapiens groups went on representing the world around them on cave walls. Their oldest rock art has been discovered in the cave of Chauvet, in France, dated to 37,000 years ago (Quiles et al. 2015). The heads and legs of 442 different animals were drawn in sequence next to each other in different sizes. Some have argued that this was the first attempt at producing a dynamic representation of observed reality, a sort of cinematic story. On an all-embracing curved "screen" of 36,000 square meters, the first IMAX show in history might have been played. This pictorial technique could render a sense of movement with the help of the flickering light of torches (Azéma and Rivère 2012). Such a show, perhaps accompanied by stories sung or told, would have elicited strong emotions in the audience, making it feel at the centre of an extraordinary spectacular event. In the caves of Altamira, in Spain, there are images with ages of more than 35,000 years. They were painted and repainted until 15,000 years ago (García-Diez et al. 2013).

The first musical instruments had been constructed from bird bones and mammoth tusks, with multiple holes carved using stone tools. Those found at the Hohle Fels site, in Germany, are 43,000 years old (Conard 2009). Nearby, in layers of the same age, archaeologists found a small headless figurine made of mammoth ivory. It had exaggerated female attributes and was designed to be worn around the neck (Conard et al. 2009). Similar evidence of symbolic representations and rituals were found in many other locations visited by Sapiens groups during the last ice age.

In those days, life was short and dangerous, for people rarely reached more than 30 years of age. When we talk about our ancestors, we must remember that, by today's standards, they were mostly young people. Playing, dancing and singing around the fire would have promoted friendships and encounters. And it would have been an excellent opportunity to tell stories and live collective experiences. Moreover, in creating the opportunity to bring people together, artistic representations could encourage the exchange of ideas and information. From then onwards, an evolutionary perspective in which individuals survive by simply adapting to environmental changes would not suffice. The fittest would become those who could better manage their social relations. For example, those who would show the sharpest minds, the highest learning skills, the bold and the beautiful, the socialites. Such a twist is key to boost the development of innovations. But it creates a tension and a stratification between the former "biological drives" and the new "cultural drives".

Indeed, almost a third of our ancestors' life took place when they were teenagers, and that is precisely the period when we are more curious and have a greater propensity to take risks. This attitude is caused by a time lag in the development of two important brain regions (Mills et al. 2014). The limbic section that incorporates hormonal messages gives a particular impetus to emotions. It develops fully after 15 years. The prefrontal region that controls these emotions develops only after the age of 25. This mismatch in mental development can become an evolutionary advantage if, on the one hand the most innovative behaviour is selected and, on the other hand, an excessive risk propensity can be contained. This last consideration is particularly important considering that, at this intermediate age, one is more inclined to learn from his peers than from the elderly.

So, there are at least three traits in Sapiens adding up age wise: a premature birth, a longer infancy and—at least for a while—an increasing tendency to venture into risky business. The conservation of juvenile characters in adults—anatomical and behavioural—is called neoteny by biologists. This condition is associated with the necessity of protecting a population of immature individuals. A safe "home" (*domus*) was needed to contain the possible excesses of youth without renouncing their contributions to new ideas. A fundamental passage was a drastic decline in social aggressiveness. By turning into a *domestic* character, gradually the "savage" was giving way to the "tamed".

6.3 Teaching and Learning

In any case, it is reasonable to assume that elders played a key role in transmitting the innovations introduced by the cultural revolution recorded in Eurasia around 40,000 years ago. To improve successful innovations, we must first master those that others have made before us. This knowledge should be modified and adapted to new challenges. Having a complex language is important to transferring information on aspects important for survival. The transmission of knowledge is also amplified by more young people and by an early familiarity with symbolic thought, a subject that we will discuss later. This expansion is therefore associated with the increase in population and the capacity to imagine new ways to apply the knowledge already acquired.

Play is very important for the application of existing knowledge to previously unexplored fields. We know how much our children appreciate the fantastic stories we tell them, and how they like to project themselves onto objects (toys) and invent situations (games) springing from their imagination. In encouraging them, we cultivate their capacity for abstraction, and therefore the possibility of generating realities transcending the observed one. That would be very useful, as we shall see, to translating these fantasies into ideal visions, tools and procedures to understand and dominate nature and to organize our social relationships. We do not know how widespread play was in the past, but we can assume that a larger number of younger people favoured this attitude. Today play keeps extending into adulthood up to senility. Eternal youth is well thought of as it favours exploratory activities even in maturity. It also inhibits neo-phobia (the hatred of novelty), an attitude that often occurs in older people. In the final parts of this work, we will offer a somewhat more pessimistic interpretation of these processes, in relation to their recent developments.

But is it enough to have an extended juvenile period and a greater longevity to justify the extraordinary human accomplishment that we attribute to our species? Surely the Sapiens have a special ability to acquire knowledge from others and to use that store of experience to devise novel solutions to life challenges. But many other species learn and innovate too. Chimps open nuts with stone hammers and extract ants from nests with sticks, when they see others doing it. Dolphins use tools to extract hidden prey from their hideaways. What is so distinctive in our capacity to teach with enough precision over the generations to build a spaceship or discover the DNA?

According to the "cultural drive hypothesis", species proficient in teaching and innovating should have larger brains. Cultural drive outlines a feedback loop between social behaviours and genetics in which better cognitive skills and bigger brains are selected by way of high-fidelity copying of others' behaviours. This process leads to improved social behaviour and enhanced dietary habits that, by way of an increasing brain volume, result in better teaching and copying. In particular, bigger brains imply, among other things, enhanced perceptual systems, better connections between senses and motion, taking another person's perspective (theory of mind), thinking about the past and the future, reasoning in hypothetical terms, better computational abilities.

The secret of human success in information transmission—it is argued—comes down to the fidelity of information exchanged among individuals and the accuracy with which learned information passes between transmitter and receiver. It has been shown that both the size of the cultural repertoire of a species and its persistence in a population increase exponentially with transmission fidelity. Once a certain threshold is reached, culture begins to increase in complexity and diversity. From then onwards, even modest numbers of novel inventions and refinements lead to massive cultural change. Cumulative culture is then possible only with accurate transmission, enhanced by the ability to generate know-how by oneself.

Under this perspective, it is argued that humans are the only living species to have passed that threshold (Laland 2017). According to the "cumulative cultural brain hypothesis", it is then possible to envisage a set of narrow conditions that favour an auto-catalytic take-off in human evolution, based on sociality and a cumulative co-evolution of population size, population structure, more sophisticated learning strategies and life history (Muthukrishna et al. 2018).

When these conditions are met, a body of adaptive information builds up over generations, leading to a selection for brains proficient in social learning and in storing and managing adaptive knowledge. When brain size reaches a certain biological limit, the increase in the amount and complexity of adaptive knowledge can continue via a division of information and ultimately a division of labour. In addition, formal education will expand the juvenile period, in which more time is devoted to learning, and create a gap between fertility and reproduction. A delayed birth of the first child will then help extend juvenile traits into adulthood. This theme keeps appearing in our reasoning under different perspectives.

Human cultural knowledge included foraging techniques, food processing, music production, toolmaking, body ornament and alteration, pigment paintings and rock or wood carving. This is the context in which language emerged in the first place. It was not the product of a sudden genetic mutation or of new brain structures. It was the result of a circular process involving a platform of ancient capacities (shared with other animals) and some modern traits. This process engaged individual learning and an accurate transmission of language. Its wider and deeper use favoured selection on hominins' brains for language-learning skills. But in order to develop fully, the co-evolution of genes and culture needed large social groups and new capacities for abstraction.

Returning to the Neanderthals, they had declined in number even before we arrived, and perhaps also had a slightly shorter average life expectancy, according to the analysis of the teeth from 70 Neanderthals from Krapina (Caspari and Sang-Hee 2004). Several evolutionary disadvantages seem to have conspired against them in their traits. If we add the observation that Neanderthals tended to live in smaller social groups, and that our arrival had further fragmented them, their disadvantages mount. Sapiens could form larger and more connected societies, by activating a virtuous cycle in the transmission and expansion of knowledge.

6.4 Trust, Gossip and Shared Beliefs

How did the Sapiens organize their daily lives within increasingly large groups? Gossip played a big role. Another important role was played by capital punishment (Wrangham 2019). The former can be interpreted as an extension of the mutual

6.4 Trust, Gossip and Shared Beliefs

grooming typical of other primates. This practice has little to do with personal hygiene but rather with cementing interpersonal bonds and exerting an influence on one's companions (Dunbar 2004). Capital punishment was instead crucial for limiting the reactive aggression we have inherited from our common ancestor with apes.

In modern humans, thanks to the use of complex language and the expansion of our mental skills, gossip became a formidable tool to cement friendships and pressure everyone to follow the rules so that others wouldn't speak badly of them. However, when we aggregate beyond a certain number, we can no longer rely on relationships based on the knowledge and trust of each other. Nor can we rely much on the reputation of the single individual. We need a more powerful social glue. In any case, what led us to increase our social numbers in the first place? The first big bands were probably formed due to population increase and alliances with other groups of Sapiens. The main reasons that come to mind were to combine and coordinate the hunting of large animals and to face rival groups. But what was the basis of these alliances? And for how long did they last?

A good basis could be agreement on a world view associated with a particular interest in working together. The construction of shared beliefs serves the first general function very well: One joins like-minded people voluntarily and with pleasure. The exchange of goods and services serves the second purpose: We are much better off if we ally with complementary people. But in the case of group alliances, in addition to sharing a common sense of life, economic and political relationships become significant.

It is well established that, in order to function, a community needs a social contract based on cooperation and castigation. However, the devotion to a collective ideal and a common interest works much better than the fear of being punished. The number of individuals that would warrant a transition from a social organization based on trust and personal ties to one based also on shared beliefs and rules is a matter of controversy. For every species, the optimal number emerges from a cost-benefit analysis of living together. Costs would derive from ecological and social competition for food, rest and reproduction. Benefits generally refer to reducing the risk of becoming prey or of being expelled from a territory, and to the possibility of sharing parental care and mutual assistance.

The evolution of morality—defined as a drive to cooperate with our next of kin—plays a key role in passing from "individual" intentionality to "joint" and then "collective" intentionality. Individual intentionality relates to the ability to achieve a particular goal and pursuing self-interest in a collective endeavour (for example when foraging for plants with mates and then eating separately). This is a typical chimp behaviour, for example. It is also at the base of the "free market" ideology of modern capitalism, when it claims that only the pursuit of self-interest can guarantee the maximum output for all and then builds up an entire doctrine based on individualism (microeconomics) to interpret, as an extension, the functioning of the society as a whole.

Joint intentionality requires heightened cooperation within a group and a focus on common goals. It needs a strong and active social selection for competent and motivated individuals who cooperate well with others and do not hog the resulting spoils of a common endeavour. Probably, this second attitude was already present in *Homo heidelbergensis* and in Neanderthals and early Sapiens (Perner and Esken 2015; Stiner et al. 2009). In this case, a sort of "enhanced morality" emerges in which—to avoid being ostracised or dying in clashes with rivals—a "me" had to be subordinate to a "we". This is the framework in which social sciences operate and is the realm of macroeconomics.

Finally, a collective intentionality, typical of modern humans, had to make additional assumptions to build up a group identity and recognise our next of kin: for example, a common language, traditional food processing, distinct social practices, aesthetic norms. Teaching one's children to do things the conventional way becomes now mandatory for survival. People internalize the proper way of treating other people and being treated with fairness (Tomasello 2016). In this latter case, modern humans not only care about what other people think of them, they are also endowed with an objective form of right and wrong. In contrast with the evolutionary approaches based on reciprocity and the management of one's reputation in the community, the term "we" now includes also what "I" think of myself. Such a morality might have emerged around 100,000 years ago and later segmented into diverse political, ethnic and religious lines with the advent of agriculture, 10,000 years ago, all the way down to the capitalism of recent days. A self-judgemental moral attitude is born.

To accommodate all of the above, and possibly other issues such as the analysis of inequality, a political economy approach is needed (Stilwell 2019). This is a frame of reference that endorses a pluralist method, a transdisciplinary inclination and an ethical orientation in evaluating economic choices. So far, it seems at odds with the prevailing political attitudes of the world and it is consistent with posthumanism, a philosophy to which we shall return in our final considerations.

From a different perspective, evolutionary psychologists have linked the increase in the volume of the frontal and parietal lobes in the different hominins to the size of the social groups they could organize (Dunbar 1995; Dunbar 2014; Dunbar et al. 2014; Powell et al. 2012). This "social brain hypothesis" was initially proposed as an explanation for the fact that primates have unusually large brains for body size compared to all other vertebrates. In primates, a quantitative relationship between brain size and social group size was observed. The cognitive demands of sociality— it was argued—place a constraint on the number of individuals that can be maintained in a coherent group. In bands of less than 50 individuals, typical of chimpanzees and pre-human hominins, one lives in the wide circle of his or her family and friends, and coordination is generated by mutual knowledge, trust and reciprocity.

According to the aforementioned theory, *Homo ergaster* could form groups of 80 individuals at most, *Homo heidelbergensis*, of 100, and Neanderthals and Sapiens, of 150. But to overcome this number (called the Dunbar number), it seems essential to use other forms of social coordination. In contrast with other mammals and birds, it seems that anthropoid primates may have generalized the bonding processes that characterize monogamous pair bonds to other non-reproductive relationships, and extended them, for example, to friendship or kinship (Dunbar 2009). Indeed, this is the premise of the "cumulative cultural brain hypothesis" discussed above. And it could be considered as the first step of a domestication process in which vertical social relations will develop according to a certain hierarchical order.

In what follows, we argue that to create a sense of belonging to a community beyond that of their acquaintances, modern humans had to maximize their skills of abstraction and generate symbols and virtual worlds. That was achieved through the development of a language capable of conveying complex information. This information belongs to three distinct levels: those relating to the reality observed in nature, those concerning ourselves and our peers within a community and finally those referring to imaginary worlds capable of uniting people who identify with the same culture. A number of other species communicate at the first two levels, but only Sapiens are able to create such a vast assortment of imaginary worlds.

Certain religious beliefs would seem particularly good at fostering cooperation with strangers and distant people to the point of becoming adaptive characteristics. The results of a recent ethnographic study based on interviews with people of different religions (Hinduism, Buddhism, Christianity, Animism and ancestor worship) are very interesting. To measure their adherence to social rules, these groups were subjected to two behavioural games. The results showed that faith in moralistic, omniscient and punitive deities was a key mechanism for building large social groups and for forming social bonds with strangers over long distances (Purzycki et al. 2016). In practice, believing in an entity that knows everything provides very efficient social coordination and avoids opportunistic behaviour due to the certainty of punishment, if not in this life, then certainly in the other life. In general, though, cultures can vary and therefore serve both to unite and to divide. This dual possibility will carry hard consequences for the complex and multicultural society of today.

6.5 Work, Leisure and Learning Today

Returning to our initial questions about the time we spend in different activities, we believe that young and old people enjoy more free time today than in the recent past. Considering the working conditions of two centuries ago, we could be right. People started working earlier and worked until they died. But does this apply to the deep past?

In truth, we cannot confirm that our working hours are reduced in passing from self-sufficiency to the division of labour. Even though by specializing in different tasks, the availability of goods and services grows, we have to contend with an increase in our needs and the necessity of developing more complex and interdependent social relations. Today, we are constantly connected and accessible through digital technology. This keeps increasing our working hours. And with this technology, we gladly replace the working hours of those still employed in railway stations, travel agencies, banks and airports, to name a few (Lambert 2017).

It seems that today the distinction between work and free time is fading away. In labelling what we do just for the fun of it a hobby, we have paved the way for countless do-it-yourself activities to which we are sometimes oblivious. Such actions bypass intermediaries and make the previous habit of buying and selling labour time partially obsolete. Even though new jobs are created, many more are lost and new competencies need time and training to build up. We operate under conditions reminiscent of those of our ancestors when being hunters and gatherers consumed all of their energy while providing fun and good tucker. Old-time working hours ought to be gradually reduced, to make everybody "earn a living", despite the overwhelming contribution of information technologies. Instead, we are induced to consume more and more. And clinging to the old paradigm of industrial production, we keep looking for jobs, even if most of them are gone and new ones are yet to be discovered.

We often believe that the organisation of the production that we experience is the best possible way to deliver what we need, no matter if that is "market capitalism", "state capitalism" or anything between. And if we don't, we must be prepared to pay a high price, as society will tend to stigmatise us, marginalise us, or worse. We call the previous two opposite systems of production by names that contradict their observed nature. When large multinational companies organise the majority of transactions through detailed managerial plans, we speak of "free markets", as if prices could really vary, by and large, according to the laws of demand and supply of atomistic agents. When masses of people work in sweatshops under conditions of deep exploitation, we accommodate for the construction of socialism, as if that ideal condition could be postponed forever to benefit future generations.

In either case, we stick to the myths that have been created around these systems, and the benefits that they are said to deliver. And following the leaders of the moment, we tend to adopt herd behaviours, and bend to the necessity of that particular system of production to reproduce itself according to yesterdays' rules. We don't even consider the idea that a political and economic domestication might be under way. Indeed, we feel offended by it, being nurtured, since an early age, with the pride to belong to that particular system.

Another well-established fact is that today we can access information on an unprecedented basis. Indeed, we are said to be living in an information society. This consideration is sometimes tempered with the idea that, though ultra-informed, very few grasp what is really going on. Indeed, our brain is shrinking in volume, since more than 40,000 years ago. Brain functioning in problem solving have improved in the meantime, thanks to better connections among our own neurons, and larger and deeper networks with people and technologies. However, our capacity to process and elaborate the incredible amount of information now available is limited, and in fact it is increasingly handed out to intelligent machines. A thorough consideration of the feedback effects of information technologies on our body and mind—in accordance with the "cultural drive hypothesis"—is highly recommended.

A final note is due in relation with the morality issue we briefly discussed above. In recent days, it seems that, at least in parts of the so called "developed world", some members of our species are reversing the trend that took several million years to develop, in organising our social relations. We briefly touched upon the shift from individual intentionality of early primates and modern chimps to the joint intentionality of some of our ancestors and contemporaries (which includes our opinion of others) all the way up to the collective intentionality of the self-judgemental Sapiens (which includes the opinion about the way—right or wrong—in which we behave).

In recent days, perhaps as a reaction to technologies that allow for wider and deeper connections on a global scale, we witness a tendency to reduce the numbers of those we consider members of our social groups of reference. By dividing humankind according to nations, ethnicity, gender, regions, townships, etc., the sense of our kinship tends to split, diversify and shrink in numbers. This tendency often implies excusing ourselves from any moral judgment on the way we treat "others". It may end up in minding our own business to the detriment of the community, and in closing ourselves in progressively smaller mental cages. This attitude is nothing more than restricting our social relations to foraging only, in a pattern that was common to our furthest predecessors. It remains to be seen to what extent this evolutionary pattern is compatible with a future of peaceful coexistence.

Chapter 7
Food for Body and Mind

Food is a modern obsession: many take refuge in vegetarian diets, organic food or "clean" and "paleo" eating, to name a few examples. Increasingly, many people tend to avoid red meat, dairy products and many sources of gluten and sugars, such as refined carbohydrates.[1] Others embrace paleo-eating, harking back to the "good old habits" of our ancestors.

But are we really required to follow diets like these to feel good and be healthy? And what diets followed the hominins in the deep time of our evolution? This is not pure historical curiosity. After all, we are not only "what we eat"; we are also "what they ate," referring to those who preceded us. And it is not only a matter of our bodies; it's also about our minds and our sociality.

In truth, despite loving vegetables, fruit and nuts, even the ancestor we had in common with the chimpanzees probably did not disdain all meat. Nor do current apes. Also, *Australopithecus afarensis* could be carnivorous. In 2010, in the geological record of more than three million years ago, bones of animals were found in Ethiopia that had probably been slaughtered by these hominins wielding rather primitive stone tools (McPherron et al. 2010). In general, however, they preferred to feed on fruit and leaves, like modern apes. This, perhaps, is why they went extinct—when climate change dramatically disrupted their habitat.

Recent analyses of the carbon isotopes in the teeth of some specimens of this genus reveal that they had tried to broaden their diet somewhat. *Australopithecus africanus*, for example, living three million years ago, regularly ate grass and sedges or, perhaps, small animals that ate these plants (Sponheimer and Lee-Thorp 1999). The last australopithecines, such as *Australopithecus sediba*, however, were resigned to a diet based on leaves and tree bark, according to the isotopic analysis

---

[1]To follow these new directions, a famous Italian manufacturer has already put on the market alternative pasta made from lentil flour, chickpeas and peas. In the United States, synthetic meat created in the laboratory is being produced to replace animal proteins, which are not very sustainable and are very polluting.

of phytoliths[2] in their dental tartar (Henry et al. 2012). On the other hand, recent biomechanical studies indicate that *sediba* had very weak jaws, unsuitable for a diet based on leathery foods. Its destiny was written (Ledogar et al. 2016).

Until recently, it was believed that, in large parts of Africa, forests had slowly and gradually reduced to the savannah. Recent analyses of sea sediment cores taken from the Gulf of Aden on the east African coast show that there were two periods in which the continent saw a rapid alternation of forestation and deforestation. The first was between 2.9 million and 2.4 million years ago, the second between 1.9 million and 1.6 million years ago. Each of these periods corresponds in turn to a significant passage in the history of human evolution. In the first one, early human forms appear, such as *Homo habilis*.[3] Another is a human species recently discovered in Ethiopia, dating back to 2.8 million years ago. It has not yet been named (Villmoare et al. 2015). In the second period appears *Homo ergaster*, the first unequivocally human species.

Very different were the evolutionary responses to these environmental changes, which caused high instability and the fragmentation of habitat (Antón et al. 2014). Some hominin species adapted with a switch to a morphology more suited to the new natural resources available (for example, through different jaws and teeth).[4] Others started experimenting with new evolutionary options, such as an increase in brain size, in particular the development of the neocortex and frontal lobes. It is as if for some groups of hominins, biological evolution had become too slow. It became useful to develop the ability to respond more quickly—through behaviour, learning and sociality—to the effects of the most rapid climate change.

The brain accounts for only 2% of body mass, but it guzzles at least 20% of the total energy consumed. A larger brain has a high energy cost, so it is favoured only if the evolutionary benefits exceed the costs. Ultimately, the evolutionary line based on the strengthening of the masticatory apparatus turned out to be a dead end. *Paranthropus* went extinct about one million years ago. In contrast, adaptation

---

[2]Phytoliths are microscopic minerals that are formed in the cells of many plants and which are enduring.

[3]*Homo habilis* had small teeth and jaws, and a brain of more than 700 mL. It was completely bipedal with short fingers and toes, unsuitable for climbing trees. It was named "Handy Man" because it was considered the inventor of stone tools, although recent discoveries question this.

[4]*Paranthropus boisei*, whose first fossil remains discovered in East Africa date back about two million years, was called "nutcracker man" due to the fact that it had flat molars, thick tooth enamel and robust jaws. It also had powerful chewing muscles to the point that to hold them, it had formed a bone crest on top of its skull. The isotopic analysis of dental enamel confirms a change in diet from fruit and leaves to berries, roots, tubers and even termites (Cerling et al. 2011). *Paranthropus robustus*, from southern Africa, was oriented toward a specialized vegetarian diet. Other hominins continued to diversify their diet. They had started doing so a long time before. Among them was *Australopithecus africanus* (Sponheimer et al. 2013).

based on the expansion of brain size continued successfully in different human species all the way down to us.[5]

This evolutionary line would have multiple advantages, the main one being linked to sociality. About 1.9 million years ago, *Homo ergaster* appeared in Africa with a brain of about 1 L. It had smaller jaw muscles and molar teeth, suggesting a diet based on soft food. These events occur downstream of a genetic mutation which 2.4 million years ago had caused the human line to lose a protein, called MYH16, which is responsible for the development of the maxillary muscles of the upper jaw (Stedman et al. 2004).

Without strong jaws, but knowing how to treat meat with stone tools and fire, *Homo ergaster* could ingest large amounts of protein-rich food and digest it relatively quickly. This further modified its anatomy. With a less bulky digestive system, the rib cage shrank even as the caloric intake of ingested food increased, as in Turkana Boy, the young *Homo ergaster* we met earlier. With smaller chewing muscles, the evolutionary constraint on the growth of the skull was relaxed. Without muscular attachments on the skullcap, there was now space for the sutures that facilitated the growth of the brain after birth. The brain of *Homo ergaster* was more than double the size of that of the first bipedal apes. And on a diet of cooked meat, it could continue to grow in later species. Probably some australopithecines ended up on the menu of these early humans, along with members of their own species.

7.1 Ritual Food

Evidence that some of our ancestors might have been cannibals comes from the Gran Dolina site at Atapuerca in Spain where unequivocal signs of butchery were found on the bones of a dozen or so archaic humans who lived 800,000 years ago (Fernández-Jalvo et al. 1999). Cannibalism continued with Neanderthals and Sapiens. It is thought that it was associated, in the latter cases, more with ritual than with nutritional needs, constituting an important hallmark of our species (Tattersall 2015). In Britain, for example, researchers found human bones engraved for clear symbolic purposes during a funeral rite based on cannibalism. They date to 17,000 years ago (Bello et al. 2017).

The precedence of ritual over alimentary requirements is confirmed by the behaviour of some populations which, until recently, practised cannibalism during funerary rites. They later abandoned it but only for health reasons. For example, cannibalism was outlawed in Papua New Guinea around the middle of last century. The legislation was not passed for ethical reasons, but to eradicate kuru, a neurological disease caused by the consumption of the deceased's brain during funerary rites.

---

[5]There are two exceptions to this trend: *Homo floresiensis* and *Homo naledi*, who both maintained a small brain until their extinction, which took place 70,000 and 250,000 years ago, respectively.

Eating can have at least two antithetical symbolic meanings—communion and aggression. The first has been elevated to a transcendent plane and accompanies rituals, which have remained very popular until today. The second has turned into a taboo: it is forbidden to eat one's fellow men, members of the great human family. In time, the rules and habits of the latter type would be extended to animals performing family functions, such as pets or companions, including dogs, cats and horses.

At one point for Sapiens, food began to be linked to the transmission of myths, religions and to the creation of institutions that regulate social life. Just think of the rituals involving animal and human sacrifices, widely documented in many cultures. This behaviour is linked to the construction of imaginary realities, the result of new and peculiar mental abilities. The evocative role of food continues in modern society, as seen in the rituals acted out daily in our homes and restaurants. Business lunches, family dinners and wedding banquets are all examples of occasions in which food is used as a means to enhance social rules.

The many dietary norms imposed by religions dictate what can be eaten when. Think of the requirement to eat only Halal or Kosher food, or when food moderation should be observed, such as during Lent or Ramadan. There are also taboos that forbid us to eat some animals, such as pork, and customs to consume others on certain occasions (for example, lamb at Easter). Food is also often a symbol in itself: Think of the bread and wine of Christianity.

Finally, the obsession of becoming leaner or more muscular can also be traced back to the myth of perfect beauty. Today, pursuing these myths can lead to serious pathologies, such as anorexia or muscular dysmorphia (a disorder that causes some men, and even some women, to develop their muscles compulsively). Even the obsession for health food could become an eating disorder, orthorexia.

7.2 Vegetarian or Carnivorous? Omnivorous

The Neanderthals have had the image as insatiable carnivores since their discovery a century and a half ago. Many animal bones were found on some sites, along with Mousterian stone tools. At first, it was hypothesized that their favourite fair was large animals, such as mammoths, elephants and woolly rhinos, but recent studies reveal that they did not turn their (big) noses up at small mammals (Smith 2014).

Yet there is a physiological limit to the amount of protein you can eat without risking toxic effects (Fiorenza et al. 2015). Recent examination of Neanderthal teeth indicates that they observed a more varied diet, including tubers and other leathery plants. Thanks to the analysis of phytoliths and starches in dental tartar, an excellent archive of dietary habits, we also know that they ate various types of vegetables and sometimes even cooked them (Henry et al. 2010). Analysis of the tartar of 50,000-year-old Neanderthal teeth from the El Sidrón site in Spain turned up no trace of protein derived from meat. In contrast, the analysis of sediments containing Neanderthal fossil faeces, again 50,000 years old, from the Spanish site of El Salt, supported the thesis of a diet based on meat (Sistiaga et al. 2014).

The DNA analysis of the dental plaque of El Sidrón Neanderthals shows that they were greedy for mushrooms, moss and pine nuts (Weyrich et al. 2017). The diet varied among the different groups scattered in Eurasia. For example, "Belgian" Neanderthals considered in the same study, while not disdaining moss, perhaps as a side dish, ate woolly rhino and wild goat meat. In the dental plaque of an El Sidrón Neanderthal was the DNA of the poplar—a tree whose bark contains salicylic acid, the analgesic ingredient of aspirin (Hardi et al. 2012). Since the individual in question had an ugly dental abscess and a bacterial intestinal infection, it follows that our cousins used the plants to self-medicate. This practice is confirmed by the Neanderthals' use of other medicinal plants, such as chamomile and yarrow. Furthermore, a modern human gene, TAS2R38, which enables us to sense bitterness to help in the detection of plant toxins, has been identified in Neandertals too (Lalueza-Fox et al. 2009).

What did Sapiens eat when they were still hunters and gatherers? Throughout this long period, our ancestors' diet was wider and more varied than ours. Despite the prestige of hunting large animals, for the most part their food was based on the patient harvesting of the earth's bounty, including plants, small animals: fish, molluscs, larvae and insects.

Before farming, thanks to a more varied diet, a few sugars and a lot of exercise, our ancestors were probably healthier and more athletic than we are today. A high infant mortality and an early disappearance of the weaker would have helped to reach this outcome. Early Sapiens certainly did not know what obesity was. Their skills had to be broader, since they could rely only on themselves and a small circle of companions. They also had to know the land, where the water was, and the techniques of hunting and collecting. They had to be aware of what was edible and what was not.

If a member of our industrial society with the modern corpus of knowledge were put up against one of our ancestors, we would probably come off worse. Our hopes of making it would increase only if we were a large group of individuals, each with different skills and competencies. We have chosen to specialize in increasingly narrow areas. In doing so, we can source from a large pool of resources, but only as members of a society. Our knowledge is much broader as a collective entity, but it is much narrower as individuals. And most of the knowledge we acquire is related to experience in a specific environment. For example, think of the early British explorers who were trying to survive in the remotest areas of Australia. It was not enough for them to imitate the Aboriginal people in getting food from local nutritious plants. In some cases, they had to know how to get rid of their poisonous effects. In fact, many expeditions ended up tragically because of a lack of contextual knowledge (Boyd 2018).

7.3 Farmers and Breeders

At the end of the last glacial period, different Sapiens started developing agriculture and selective breeding of some animals for food. Sedentarization increased mothers' fertility and reproductive success. This would have accelerated the previous population growth, even though wars, famines and pandemics put the brakes on it. It is thought that the net effect was positive for the farmers (Page et al. 2016), and that almost all of the previous hunter-gatherer populations were replaced by those who devoted themselves to agriculture and breeding. In reality, however, this thesis is at odds with other investigations.

The long-term global population growth in the Holocene does not show substantial differences between the two types of societies (Zahid et al. 2016). Deploying radiocarbon dating on objects representative of human activities found on different continents as a proxy of the demographic trend, and considering short-term oscillations, hunter-gatherer societies do not seem to show a demographic slowdown compared with their contemporary farmers and breeders.

If the latter had taken over the former, then the reasons should be sought elsewhere, a higher population density of the farmers, for example. We know that the largest permanent concentrations, first in the villages and then in the cities, correspond to the advent of agriculture. And then, in the past 10,000 years, the general picture might have been roughly the following. On the one hand, we would have had many small groups of hunter-gatherers, mobile and scattered all over the planet. On the other hand, we would find few groups of farmers, much more stable and concentrated, and able to enjoy the advantages of the division of labour of larger communities and the consequent economies of scale in collective production. The domestication of the environment would then go hand in hand with the development of pro-social attitudes. A circular and cumulative process of domestication of both nature and individuals was activated. Eventually agriculturalists would replace and assimilate hunters and gatherers in a growing number of geographical areas.

This success story has a number of drawbacks. Focusing on quality, even though agriculture and livestock increase the amount of food available to everyone, only the richest can increase the variety of their diet. Even today, most of humanity lives mainly on a few cereals. In general, the diet of farmers and breeders opened the way to new diseases: from caries to periodontitis, from infectious diseases to iron and other deficiencies, as well as to the spread of intestinal parasites. And industrial breeding of animals for food is now putting the environment at risk as a whole, via extreme carbon dioxide emissions. The clearing of large forests for intensive agricultural activities multiplies these effects. Again, all these difficulties, in requiring new cultural innovations and countervailing measures, strengthen social interdependence.

Setting environmental considerations apart, after 10,000 years of this experience, we begin to consider the option of feeding on smaller and more abundant animals, including species to which part of humanity responds with disgust, such as insects and larvae. These species are still part of the diet of many populations, especially in Asia and the tropics. To meet the tastes of the industrialized countries, projects are under way to commercialize insects through the production of flours from which "Westerners" can make dishes more to their taste. The Danish company, Enorm, already produces crackers with larvae flour and cicadas. Dehydrated insects in the form of biscuits are beginning to be distributed in European supermarkets. In 2017, Switzerland launched flour-based meatballs with *Tenebrio molitor* (mealworm larvae) in its supermarkets. Meanwhile, China has started a program to investigate the possibility of including some larvae (such as the silkworm) in food to be cultivated during the intergenerational space travel of the future. (Jones 2015).

Going back to diets, and the evolution of social organisms, we have said that the increase in brain size, which started two million years ago, helped the humans of the time find original and innovative solutions in order to survive. These solutions were mainly cultural, being based on the invention and use of increasingly efficient lithic tools and on the use of fire to cook. This knowledge had to be stored, organized and then transmitted to future generations. To do this, a bigger brain was needed. Then, inventing new tricks to treat food (extraction, processing, combinations of ingredients to eliminate toxic or harmful effects) the knowledge coalesced in a collective heritage. An even bigger brain was needed to handle all this information, and possibly involve an increasing number of individuals. Recent evolutionary neuroscience claims that this process lead to a "unique human technological niche rooted in a shared primate heritage of visuospatial coordination and dexterous manipulation" (Stout et al. 2017).

Once the process of interaction between biological and cultural evolution began, it tended to feed off itself and accelerate, as in an autocatalytic process (Laland 2017). It could be limited only by biomechanical and physiological constraints. Between 300,000 and 100,000 years ago, humans reached this physiological limit, with a brain in adulthood of about 1.5 L. With Sapiens, the process that connected (circularly) biological and cultural evolution did not stop but became paroxysmal. This generated a social organism which was also a repository of common information on survival. It was not our individual intelligence that helped us survive and conquer the planet. It was our collective intelligence.

With the division of labour, probably first between man and woman, and then within a larger social group, it was possible to decentralize many functions that were once the preserve of individuals. Later, modern Sapiens found other strategies to manage the surplus of thought that they could generate collectively. And they went on transmitting and reproducing it. Focusing on their diet, how much did the new food options contribute to empowering the Sapiens brain and improving their sociality?

The conditions of the last glacial period were very difficult for everyone—for the different human species that populated the planet and for the animals and plants they ate. While the cold and unstable climate of Eurasia proved a challenge for Neanderthals and Denisovans, the waves of drought in Africa made the savannah increasingly inhospitable to Sapiens. A world of plenty was rapidly turning into a world of scarcity. Recent research reveals that the coastal environments of southern Africa were the salvation for our direct ancestors. In reaching the shores to escape drought and famine, these Sapiens began to alter their diet radically, and continued to do so even in their wanderings outside Africa when they hugged the coast. Even today, a good part of humanity—more than 75% – is concentrated along the coasts. The remaining 25% live mostly near inland water resources, leaving large empty spaces inside the continents.

This love for the coastal areas allowed Sapiens to expand significantly the variety of its food, with fish, molluscs, birds, bird eggs, turtles, marine mammals and a wide range of aquatic and terrestrial plants (algae and coconut) available in all seasons. It was therefore possible to enrich the diet with substances crucial to the development of some mental abilities, such as selenium, iodine, zinc and many other minerals that we now know are associated with an increase in IQ, verbal skills and foetal health during pregnancy when most brain cells develop.

These substances reduce the incidence of mental illness and of cognitive disabilities. In particular, it is known that long-chain polyunsaturated fats, typical of a diet based on the main coastal resources, help strengthen brain functions (Smail 2007). But are these elements sufficient to argue that better cognitive abilities would derive from coastal nutrition, and that they would play a key role in cultural evolution and the formation of increasingly complex societies? So far, the evidence is circumstantial.

We know that the South African coastal hunters and gatherers of 80,000 years ago, in becoming more sedentary, had also greatly increased their cultural complexity and that there already existed a certain division of labour. When the population's density increases, the social organization tends to assume a hierarchical structure and social networks extend over long distances (Marean 2017). Other major traits have probably emerged in our individual and social behaviour. Indeed, we may still carry some of these traits today.

7.4 Us and Them

Often, we are hostile to "others." But who are they? And how do we distinguish between them and ourselves? In most cases these are people considered different because they speak incomprehensible languages and have views and habits we consider bizarre. We can become particularly aggressive when we fear that they will wreak havoc on our economy and society. Annoying differences might involve skin colour, physiognomy, social status, sexual preferences, food preferences, political ideologies, religious beliefs and more. The history of humans is peppered with bitter conflicts between groups considering each other different. Even children tend to marginalize those they consider physically or psychologically different. Sometimes, this escalates to real persecution.

Conversely, we also have a high propensity for cooperation and empathy, helping those in need or teaming up with those who share certain characteristics with ourselves. However, the boundary between "us" and "them" changes continuously: It can centre on its own bell tower, its own nation or its own ethnic group. Until now, it has been thought that these distinctions derived only from cultural, social and economic backgrounds, and that they could be mitigated if they went too far. This goal would be much more difficult to realize, however, if this relational structure, based on variable geometry, did not depend only on our good will but was also linked to how our brain works. In other words, is it possible that we are bound to make arbitrary distinctions between us and them, and to what advantage?

According to recent hypotheses (Marean 2017), the inclusion of some individuals in our community, with the exclusion of others, is linked precisely to the affirmation of a lifestyle based on the exploitation of plentiful and concentrated resources, such as the marine bonanza Sapiens found on the coast of Africa, while escaping the continental drought of the late Pleistocene. The oldest economies of this type have been identified in southern Africa. When, from being scarce and undependable throughout the continent, food resources turned into stable and predictable, along the coast, they became particularly valuable. An incentive to create large defence groups soon emerged.

Inland on the savannah, it was not particularly helpful to form large groups since food was scattered, mobile, and unpredictable. Alliances were probably limited to certain occasions of big game hunting, while the cost of overseeing vast territories to keep others out, was too high a price to pay. But on the coast, it is estimated that an individual could easily collect, in 1 h, molluscs and other marine resources equivalent to 5000 calories. This small prey could not escape and was therefore coveted. There were 23 h left for rest, leisure and socializing. Similar situations could occur near lakes and lagoons.

As it was now appropriate to invest in the defence of these resources, part of that free time could be used to sharpen ingenuity and acquire new, more efficient weapons. The first private property, albeit collectively owned, would have been born (Marean 2015, 2017). Once family members and acquaintances reached a certain number, it was also necessary to work out how to recognize the ones with exclusive rights to the resources. It would have become useful to identify them through signs, such as drawings, or incisions on their skin or particular hairstyles. At the time, the Sapiens already had an imaginative brain. Soon after, they would also have mental abilities favourable to the formation of large groups.

It remains to be established how these mechanisms would have been triggered and how they would have been handed down. If selection processes based on genetic mutations were necessary, this evolutionary phase could have been quite long. But adaptation to the new environmental conditions and the establishment of new traits in subsequent generations could point to other much faster mechanisms.

7.5 Turning Our Genes On and Off

Epigenetic mechanisms allow genes to be switched on and off relatively quickly. This feature applies to gene expression both to determine our behaviour and our physical traits. The events of our life, and the environment can control the "switches" that condition the expressions of the genes connected to our neural networks and our physiological circuits. In this way, we produce inheritable behavioural characteristics. Epigenetic changes do not alter the genome of individuals. They simply select which genes are actually expressed. For example, by turning off some, and turning on others, they determine how the body responds to environmental factors, such as toxic agents, stress, nutrition, parental care, substance abuse and pathogens.

This was a concept already formulated by Darwin himself (Darwin 1859)[6] when, alongside natural selection, he spoke of the "conditions of existence" to which the various species were subjected and to which he ascribed great importance. But there were too many similarities with the previous (presumably, wrong) thesis of the French biologist, Jean- Baptiste Lamarck for Darwin's view to find consensus among his successors (Marsh 2007). This idea was soon forgotten, only to re-emerge, in other respects in the most recent years.

Let's see how the genetic and cultural processes developing in the coastal hunter and gatherer societies of southern Africa in the late Pleistocene might have interacted. The conditions existed for the selection of cooperative behaviour among members of the same group. However, aggressive attitudes toward different groups also were selected. Both of these behaviour patterns, if conveyed through epigenetic mechanisms, could take quite a short time to be passed down to the following generations. At a certain point, the first truly territorial wars would have broken out.

So far, the oldest evidence for a massacre dates back 10,000 years to a site near Lake Turkana in Kenya (Mirazón Lahr et al. 2016). The site was then a lagoon rich in aquatic resources. The bone remains are those of almost thirty men, women and children with signs of violent wounding: perforated skulls, broken ribs and various multiple fractures. In some cases, the remains of the deadly obsidian blades wielded in the massacre are still embedded in the bones of the victims. There is more recent evidence that war and violence were part of the lives of our hunter-gatherer ancestors (Schwitalla et al. 2014). It is plausible to hypothesize that such conflicting behaviours have occurred even in previous times, in the presence of abundant, predictable and concentrated resources, surrounded by vast barren areas.

The interest in controlling the territory also would have had a significant effect: It would have given a strategic advantage to the more combative, better-trained, and better-armed groups, marking the start of an arms race. Over time, these differences could be reinforced and stabilized through linguistic variations; this would have erected a communications barrier between groups. But even the concentration of obvious physical characteristics, such as skin colour, the shape of the eyes or nose, and the structure of the body, would help distinguish between us and them.

If these hypotheses prove correct, for thousands of years individuals would have been selected with brain connections favourable to collaborating with "similar" people and to competing with "different" people. This attitude would be activated by epigenetic mechanisms selected through cultural and environmental factors. To become a collective intelligence, we would then have reached a compromise: on the one hand, we could have been very empathetic among our own with indisputable

---

[6]His insights had no connection with genetics. Darwin was unaware that an Austrian monk, Gregor Johann Mendel, in 1865 published a study on the hereditary characters of peas, anticipating modern genetics.

benefits, but on the other we would have promised ourselves to continually wage war on others.

We could also use these parameters to explain our current propensity to create societies based on arbitrary categories of homogeneity and differences based on various kinds of signs adapted to the values of today's world. This would also explain the ambivalence and variability of our current cooperative and competitive nature. In fact, the areas to be defended against the "others" have only proliferated in the past 10,000 years, as the mass murders of recent history attest.

Today, there are other resources to be defended or conquered in addition to concentrated and abundant food (but only for a few). They include the wealth accumulated and badly distributed, fossil fuels, drinkable water, access to markets, technologies, even citizens' rights and pleasant lifestyles. The reasons for conflict abound. But if the empathetic component of our social behaviour continues in counterpoint to our antagonism and aggressiveness, it is difficult to see a solution that suits everyone in a world of increasingly scarce resources to be divided between the needy and the ultra-rich. A self-destructive device is implanted in the evolution of our species. The same flexibility that helped us adjust to the variance and variability of the environment (natural and social) can turn into increasingly violent behaviours. Unless we find a way to reduce our social aggressiveness.

We have already made some reference to such a possibility when speaking of self-domestication and will return to the subject in the final parts of this work. But we can anticipate that, when the first hierarchical social structures are formed, empathy will be contained and limited to societal members, and each will respond first of all to his authority figure.

We started speaking of food and dietary habits and ended up stressing the capacity to respond quickly to environmental challenges as members of a social organism. Another cumulative self-reinforcing loop between biology and culture—mediated by symbolism—appears before our eyes. Is there an evolutionary advantage in being flexible and versatile regarding who our friends and foes are? Maybe there is. But only up to a certain point. Then things can get nasty, and we need to curb the influences that take us there.

Chapter 8
Diseases and Grief

When we meet someone, the first exchange of pleasantries almost always goes around one's state of wellbeing. But if someone starts to itemize his or her ailments, we immediately get annoyed. We just wanted to start a conversation. Good health should be a natural thing, and illness an anomaly. Actually, being and keeping healthy is a constant battle. We walk every day on a steep ridge from which at least two great forces conspire to make us fall: the genetic heritage of our ancestors, including those from the deep past, and the adequacy of our lifestyle in our current environment. Adverse genetic traits should have been lost in the long run. But this is not always the case. Thanks to our new life as a social organism, we can preserve our individual weaknesses. In spite of all the different infirmities afflicting our body (and our mind), we live an increasingly longer life.

We have mentioned the interbreeding of Sapiens and Neanderthals that occurred on various occasions in the past. Today, the traces of those encounters are very diluted and range from 2 to 4% of the DNA of non-African populations. How can we single them out? The regions of our genome that derive from Neanderthals contain some hundreds of thousands of bases on 3.2 billion base pairs of our genome. But Neanderthal DNA is different from ours at specific points on the genome, meaning that these traits can be used to identify their genetic contribution.

Considering specific genes related to our health, some scientists have discovered that the DNA of Neanderthal origin could be associated with the predisposition to diseases, such as diabetes, Crohn's disease (an intestinal inflammation), lupus (an autoimmune pathology), cirrhosis of the liver and addictive behaviour (Sankararaman et al. 2014).

Another study correlates the diseases of 28,000 Americans of European descent with their genetic variants of Neanderthal origin. It showed that the genetic traits we inherited from Neanderthals increased the risk of various dermatological, immunological and psychiatric diseases (Simonti et al. 2016).

This does not mean that the Neanderthals and the late Pleistocene Sapiens necessarily suffered from these diseases. In fact, certain Neanderthal genes must have been beneficial to the early Sapiens who arrived in Eurasia. For example, those that promoted higher keratin levels, which provided them with more waterproof, cold-resistant skin, hair and nails. Keratin also defends against pathogenic organisms, an effect that helps fight wound infections. Some inheritances, though, have become a risk to present-day humans, particularly those living in industrialized societies. Genes that strengthen our immune system, for example, were probably very useful during the last glacial age, but today they increase the risks of inflammation and allergies. Hyper-coagulation was a crucial trait for the Palaeolithic lifestyle when wounds had to heal quickly, but today it can increase the likelihood of stroke and other diseases caused by reduced blood flow.

A further study was carried out on today's 112,000 Britons to see how genetic variants of Neanderthal origin influenced some traits of current European people. There are effects on one's tanning capacity, mood and other traits related to the circadian rhythm (Danneman and Kelso 2017). These genetic variants were probably selected for in Neanderthals in response to the low solar radiation levels in Eurasia and then usefully transmitted to sun lover Sapiens. In contrast, a recent analysis of the genome of the remains of a Neanderthal woman found in the cave of Vindija, Croatia, demonstrated that there were 16 genetic variants that could be linked to our current problems of high cholesterol, eating disorders, predisposition to accumulate fat in the abdomen and responses to psychotropic drugs (Prüfer et al. 2017). Modern life seems inconsistent with a "selection for the fittest" generated in a harsher and wilder environment.

There are areas of our genome, however, that are totally devoid of Neanderthal genes. This means that the introduction of other possible changes to our genetic heritage, if detrimental to survival, have probably already been removed by natural selection. Areas without Neanderthal contribution include, for example, genes related to the male reproductive system. It would seem that the genetic inheritance of the Neanderthals has made male Sapiens less fertile, lowering their likelihood of generating hybrids. On the other hand, when Sapiens and Neanderthals intersected, they were already at the limit of their biological compatibility and therefore hybridization was the last chance for the Neanderthals to transmit their genes (Sankararaman et al. 2014).

8.1 Disease and Therapies from the Past

How was the health of hominins in general? It seems that various diseases, from tumours to dental problems, afflicted both our Sapiens ancestors and Neanderthals, along with other human species. For example, the teeth of an "Algerian" *Homo heidelbergensis* of 700,000 years ago (Zanolli and Mazurier 2013) showed that caries had damaged both the enamel and the dentine. Even the Neanderthals suffered from caries and periodontal diseases (Lebel and Trinkaus 2001), as well as from more serious diseases. For example, a tumour has been identified in a Neanderthal found in Krapina, a Croatian site (Monge et al. 2013).

As for therapies, a dental operation was discovered on the decayed teeth of a Sapiens who lived in Tuscany about 13,000 years ago (Oxilia et al. 2015). Bitumen had been mixed with medicinal plants, presumably as an antiseptic. Since we were still in the Palaeolithic, before European agriculture, it is difficult to pinpoint the reason for these dental pathologies and the burgeoning treatments. Perhaps it can be put down to eating the first cereals or sweeteners, such as honey, brought by migrants from the Middle East.

The deterioration of our masticatory apparatus would become dramatic after the spread of agriculture, and this would lead to a refinement of dental therapies. The first drilling of the teeth (with flint tips) can be traced back to 9000 years ago, in present-day Pakistan (Coppa et al. 2006). The oldest known dental filling dates back 6500 years (Bernardini et al. 2012). It is ironic that the mandible with this filling, found in Istria at the beginning of the last century, could be admired in the Natural History Museum of Trieste since 1911 without anybody noticing its significance. Only recently was this detail revealed and dated. Beeswax was used to treat a damaged canine tooth. Since beeswax is an organic material, it was possible to match the radiocarbon age of the tooth with that of the filling, and write the ancient dentist into history.

What do we know in general about healthcare during the last ice age? The fossil evidence is scarce. We know that a Neanderthal of about 70,000 years ago, whose remains were found in the cave of Shanidar in the region of present-day Kurdistan, had long survived but was not self-sufficient (Trinkaus 1983). And this was not the only case. This suggests that they could enjoy prolonged assistance from other members of their group. Even the traces of the aforementioned medicinal plants in the Neanderthal diet seem to confirm the existence of cures and caretakers in very remote times.

The fossil remains in our possession tell us about their owner and his or her entourage. The collection of information on the various healing practices could help elucidate when and where relationships of trust and empathy began to form between members of a group.

We will see later that this is central to identifying an important passage of our evolutionary path, one that led us to develop symbolic thought and become the social creature we are today. It seems established that the different therapeutic practices are particularly effective beyond the active ingredients of the remedy when the patient is convinced of the efficacy of the cure and believes in the ability of its administrator to rescue him from suffering or death.

The placebo effect has been strongly re-evaluated in recent years by mainstream medicine and others (Benedetti 2014). In many cases, it is a real therapy capable of healing the patient. Examples of practitioners include the shamans, still present in some populations. Some modern-day doctors also prescribe placebos with undeniable therapeutic success. It is difficult to overstate the importance of our mental attitude to healing. Typical of modern Sapiens, this feature will have many other applications in which we think that what we believe in will eventually come to fruition. Yet there is a drawback. This characteristic can also turn against us if we believe we are being subjected to useless or harmful treatments. Or if we reject treatment, overestimating its dangers compared with its benefits. In other words, by believing that some therapy works (placebo) or damages us (nocebo), we can get the expected results, but nothing assures us that this will always be the case.

From a slightly different point of view, it has been observed that many of the treatments practiced in the past were experimental and totally pointless. Yet they could work. Why? It has been argued that sometimes it is not the active ingredient that cures but the attention of those who administer it. To overcome minor ailments, we might need "superfluous" care. This is because, as primates, we frequently used mutual grooming, a practice that is much more about bonds than mere cleanliness (Morris 1967). When we don't feel those bonds, we may get sick. To recover, it would be enough to be looked after. Health is not always personal. It is often a social thing. Empathy is thus a component of good medicine.

The previous argument can be reversed and people can get sick by social contact. When people come from other continents we can contract diseases for which we have lost, or have never had, appropriate defence systems. In our recent history, many cases confirm these fears. And indeed, as soon as the suspicion of a pandemic arises, we tend to develop safety and prevention measures that sometimes prove to be wrong or exaggerated, even if not entirely unjustified. Since 1980, in every decade the cases of transmission of many contagious diseases have doubled (both between animals and humans and between humans and humans) (Smith et al. 2014).

In our journeys of the past, we have transmitted diseases that have battered or decimated entire populations lacking resistance to them. The past 500 years are rich in documents that blame European colonization for such carnage in the Americas and the Pacific. And today we fear migrations of people and animals for similar reasons. Without detracting from the amount of evidence, it must be said that the spread of many diseases is a complicated story. A cold case of our deep past is illuminating in this respect.

Until recently, it was thought, for example, that tuberculosis had arrived in the Americas for the first time with Europeans at the time of Columbus. However, some archaeological studies have shown that it was there before. Furthermore, a recent analysis of the genomes of 259 *Mycobacterium tuberculosis* strains has led to the conclusion that all the existing tubercle bacilli evolved from a common African ancestor of 70,000 years ago. They have travelled with us, first in our global migrations and then in all subsequent demographic expansions (Comas et al. 2013).

In another study (Bos et al. 2014), the origin of tuberculosis in the Americas was reconstructed by analysing the *M. tuberculosis* genome in 3000-year-old skeletons discovered in Peru. They had deformed spines and other signs of the disease. It seems that the DNA of their bacillus is closer to that found in seals and sea lions than to present-day humans. The common ancestor of the strain found in these ancient Peruvians, and in the animals mentioned, dates back to 6000 years ago. That is long before Europeans sailed across the Atlantic. Where did it come from?

According to some scholars, these strains of *Mycobacterium* also evolved in Africa, where today they have, like us, a greater genetic diversity. Then they infected the coastal populations of seals and sea lions, which in turn spread them in the southern hemisphere to South America. This is how Peruvian seal fishermen, going after the animals' flesh and skins, were infected. Then new strains were brought from Europe some centuries later, revamping the spread of the disease in the Americas. The reconstruction of these events leads us to be wary of simplistic answers in attributing blame for a global contagion to this or that specific event and supports a holistic approach in health care and illness prevention.

8.2 Diseases of the Present: A Possible Mismatch

A brief recap is perhaps useful at this point. Indeed, a cost-benefit analysis of our evolutionary traits in response to environmental changes does not always turn out positive, especially for contemporary humans. The negative effects accelerated when we became a complex social organism and started living in an increasingly protective cultural niche, poorly suited for our biology. We entered into a self-reinforcing cycle where cultural adaptation increased the mismatch between our body features and the environment, which in turn increased the need for cultural adaptation.

In assuming the upright position, for example, we are exposed to hernias, haemorrhoids and varicose veins, and we put our spines under stress. Such effects might be negligible in youth but increase with aging. By resorting to running, we also put our joints under pressure. In order to talk, we lowered the trachea and then lost the ability to swallow and breathe at the same time, risking suffocation during meals. By reducing the variety of our diet, we are exposed to many diseases, lowering our immune barriers. By increasing the consumption of sugars and cereals, we are vulnerable to tooth decay, diabetes and obesity. In becoming sedentary, we concentrate in large urban areas where we are more susceptible to epidemics, infections and allergies. By spending long hours reading and watching screens at close range, we become increasingly short-sighted. By eating unbalanced diets and doing little exercise, we accumulate excess fats and clog our arteries. We also suffer from sleep disorders, which damage our physical well-being and our mental abilities.

In a nutshell, although average life expectancy has lengthened, health has not always improved. Certainly, modern medicine provides all sorts of drugs to help us cope with the deterioration of our health and our increased weaknesses in an ever-protected environment. Numerous diseases have been eradicated, but others have emerged. Contemporary humans are capable of adapting to the urban and technological environment of our day and endure pollution, but then we pay a price with an increase in anxiety and depression and a deterioration in the quality of life. Often, people cope by abusing drugs.

It has also been shown that we are becoming resistant to many treatments for infection due to a reckless use of antibiotics in ourselves and in the animals we breed for food. Furthermore, starting from correct hygiene concerns, we end up being obsessed with the sterilization of what we ingest, and so we also eliminate many species of friendly bacteria that help us survive. It is proven that the good functioning of our biome is not only connected to our physical well-being but also to our mental health. By eating food in which other forms of life remain (and therefore food that is not sterilized during various industrial processes), we are also more serene and balanced.

In short, we could be victims of an evolutionary mismatch that leads us to regress, holding back genes that limit our physical and psychic efficiency. Our cultural evolution is faster than our biological evolution and the progress of medicine and implant technologies try to fill the gap. However, this is helpful only in the short run. Soon, and perhaps even now, many of us would not be able to survive without industrial food, shoes, glasses, clothes, artificial implants, drugs and even smartphones and tablets. Certainly, by circumventing natural selection, we increase our present chances of survival. But in the long run, though more resistant and long-lived, we become anatomically more vulnerable, especially if we take a moral stand in favour of the weakest. In sum, Sapiens have entered an irreversible path of co-dependence with culture and large population numbers that, started as biologically driven, is now strongly technology driven.

As for any species, our physical appearance is the result of all of the evolutionary compromises we had to make in the interaction with the environment. It is a DIY bricolage that we have inherited from fish, amphibians, mammals and primates in which solutions adapted from deep time and distant environments are recycled, some still useful, others less so. We progressively add to them new elements we designed. We are hybridising with new intelligent creatures: those of our own making. They are now influencing the evolution of our body and mind, making some think that we—eventually—shall become gods (Harari 2017). Before then, we must accept that people pass away. So far, the only thing we can do is to honour them.

8.3 Funerary Rites

We attach enormous importance to death, marking it with very elaborate rites. Over the millennia we have built complex belief systems, all based on the notion of our other-worldly existence. Our ego is so huge that while we have no problem in perceiving the absence of life before birth, we can't believe in the absence of life after death. And of course, that applies only to ourselves and to our species.

The deep roots of our funerary behaviour have been studied extensively in past hominins and even in non-human primates (Pettitt 2010). A site with the bones of several australopithecines, discovered in the 1970s in Kenya—"the first family", according to some paleoanthropologists—is an example of "structured abandonment", that is a situation in which the dead bodies had been placed in a certain place, more than 3 million years ago, for a specific reason. In chimpanzees, the attitudes to their deceased ranges from morbid attention to the lifeless body to moving the remains to specific places. Occasionally, chimps dismember and consume the corpse (Cronos compulsions).

In the above-mentioned study, it is argued that the complexity of human behaviour towards the dead can be interpreted as a recent codification, ritualization and symbolization of practices that originally had no ritual or transcendent characteristics, and which are also observed in other species. Giving special care to the death of one's own seems also to be a Neanderthal practice, although we cannot claim that burials were widespread. Funerary practices are widely documented in modern Sapiens.

Australia has the earliest examples of funeral ceremonies. In 1968, from the dunes that now mark the southern shores of Lake Mungo, emerged the remains of a Sapiens woman of about 40,000 years ago. An elder of the Muthy-Muthy, the local Aboriginal people, told the discoverer: "You did not find Mungo Lady: she found you". In fact, her appearance was very timely, coinciding with the battles for rights of the first Australians. Her burial was very elaborate. The body was first cremated, then the remaining bones were shattered and finally covered with a layer of ochre dust.

Funeral ceremonies became a constant in the behaviour of humans that spread throughout the continents. A famous burial is of the Principe delle Arene Candide, in north-western Italy, dating back 24,000 years. The body was of a young man of 15 (Pettitt et al. 2003). He was named "prince" because of the wealth of objects placed around him. His funerary equipment included four tools called *bâtons de commandement* made of elk antler, a headdress of perforated shells and deer teeth, a crown of the same elements placed on his chest, a flint weapon held in his hands, and amulets and other objects worked in mammoth's ivory. The body and objects were arranged on a bed covered with red ochre dust, as in the case of Mungo Lady.

This young man was probably killed by *Ursus spelaeus*, a bear that could have weighed half a tonne. These animals had shared their caves, on and off, first with the Neanderthals and then with Sapiens. It is thought that Neanderthals faced them directly in body combat in a bid to oust them. The prince and his companions threw weapons instead. But this time the bear got the better of him. The victim's body was carefully recomposed, using yellow ochre material to rebuild part of a missing jaw, which the animal had yanked out, together with his shoulder.

In Italy there are many other sites of ritual burials. In a cave of Ostuni, a woman in the seventh month of pregnancy was buried 28,000 years ago with her headdress and shell bracelets. Her body was surrounded by horse teeth and engraved pieces of flint (Nava et al. 2017), a tribute that adds to the fond care with which her body was disposed. In Austria, two newborns were buried, decorated with honours, at a site dated to 27,000 years ago (Einwögerer et al. 2006). The bodies (probably twins) were covered with ochre powder, decorated with dozens of mammoth ivory beads and protected with a scapula from the same animal.

Even more elaborate funerary practices are found in other areas of Eurasia. In Sunghir, Russia, 34,000-year-old burials were found of an adult and two children (a male and a female) (Formicola and Buzhilova 2004). They had ornaments made of thousands of mammoth ivory grains, and headgear and belts decorated with fox teeth and other jewels, all covered with the inevitable ochre powder. The work needed to prepare their sepulchres was enormous and required the contribution of many people. As it takes at least 40 min for a good craftsman to make a single pearl of mammoth, it follows that about 3 years of work were needed to adorn the two young people with their 10,000 grains of ivory alone, not to mention the time taken to make the other ornaments. In Beringia are the remains of the cremation of a 3-year-old child dating back to about 12,000 years ago. Digging under its ashes, researchers discovered the remains of another baby and a foetus buried next to the decorated horns of animals and spearheads (Potter et al. 2014).

So far in Eurasia, a hundred elaborate burials have been discovered, dating back between 45,000 and 10,000 years ago (Riel-Salvatore and Gravel-Miguel 2013). For the most part, they contain many objects of daily life. Only a few (about ten) are particularly rich and extraordinary. The human remains mainly comprise males, but also a few females and children (Vanhaeren and d'Errico 2001, 2005). We deduce that individuals or families more illustrious than others already existed at the time.

To speculate about who made these burials and why is often very difficult and risky. Some have argued that the study of the burials and the origin of the cult of the dead in humans takes place far too often out of context. They say it pays scant attention to other objects of the same culture. In the case of the Upper Palaeolithic Gravettian tombs, for example, it would be particularly interesting to link the objects found next to the dearly departed with the broader framework of that culture: weaving, textile dying, ceramics technologies, portable art and personal ornaments (Nowell and Pettitt 2012).

8.4 The First Hierarchical Societies

To consider these points from a different perspective, we propose some general hypothetical reflections on what kind of society could be associated with the richest and most complex funeral ceremonies. First of all, there must have been a surplus of resources. The amount of work needed to produce the most sophisticated artefacts certainly required the sustenance of the people involved. This situation is consistent with the plentiful mammoth-based resources in Eurasia in the corresponding period. Large deposits of fangs and bones from hundreds of specimens of these animals have been discovered (Shipman 2015). It is uncertain whether they were slaughtered or died *en masse* from natural causes. But none of these mega-cemeteries, which date to between 40,000 and 15,000 years ago, preceded our arrival.

We do know, though, that during this period an economy based on mammoth hunting emerged. Different parts of this animal were used in the most disparate applications, including food, clothing and the construction of tools for work and combat. Mammoth parts were also used for personal ornaments, musical instruments, symbolic objects and even cooking and heating. There were also villages of shelters built from the fangs and skins of these animals, in which people lived and carried out various specialized processes. It was possible to reconstruct the entire settlement with lodgings, pits for storage and garbage, hearths, tool construction spaces, working areas, dumping sites, butchering zones and their connections (Iakovleva et al. 2012).

Economic conditions during the Gravettian period promoted, particularly in eastern Europe, an increase in population density, sedentism and changes in subsistence practices (Nowell and Pettitt 2012). A complex, hierarchical system was being built and mortuary ceremonies were an important part of the new socio-economic order.

We know that the Neanderthals also buried their dead. Several burials have been identified in Italy, France, the Caucasus and the Middle East (Rendu et al. 2014). Although not as elaborate as the ones described above, it seems that they were nevertheless performed with rituals. For example, in the tomb of a Neanderthal child found in the cave of Dederiyeh, 400 km from Damascus, the skeleton, very well preserved, shows that the child had been buried with his arms extended and his legs bent. A rectangular limestone slab was put on his skull and a small triangular piece of flint on his heart. In the case of another Neanderthal buried in the cave of Shanidar, in the current Iraq, a large amount of pollen suggests that flowers were lain. A Neanderthal child discovered in Uzbekistan was found buried surrounded by ibex horns.

A Neanderthal skull, discovered in 1939 inside the Guattari cave, south of Rome, with an age of about 50,000 years is the cause of a dispute. According to the discoverer, Alberto Carlo Blanc, the skull was placed at the centre of a circle of stones and animal bones. An enlargement of the occipital hole, which seemed intentional, led him to suggest that this was a ritual killing carried out to extract the brain and eat it. Anthropologists, however, attribute the hole to a hyena bite. They say the place of discovery was the animal's den. This can explain animal bones in the surrounding area.

Yet, other examples in several European archaeological sites testify to a certain taste of our cousins for their dead. A recent study of about 100 Neanderthal remains, dating back to 40,000–45,000 years ago, found in the cave of Goyet, in Belgium, reveals unequivocal signs of butchering on them (Rougier et al. 2016). Since Sapiens had not yet arrived in those places, some cannibalistic practices among Neanderthals seem to be confirmed. It remains to be seen whether this behaviour was due to some sort of "survival cannibalism" deriving from the desperate environmental and isolation conditions that they had to endure before the arrival of *Homo sapiens*. Otherwise, one could imagine the emergence of symbolic behaviour for this species as well, and therefore the possibility of a cannibalism ritual.

8.5 Bones, Tombs and Relics

What can we say about our relationship to the dead in more recent times? We know from experience that the loss of a loved one creates a vacuum that we fill with our mourning. We find solace in imagining communication with this person through some ritual. But how can we justify our attachment to the remains of people we have never known, and who may have lived many millennia before us? In fact, it is precisely this last attitude that we want to talk about because our feelings towards people of the past help us to lay the foundations for understanding many things about ourselves and the societies we tend to form. Attention to the ancient bones obviously has different motivations.

A primary reason for being interested in these remains is of an ethnic/cultural nature and is linked to the defence of one's identity as a people, something perceived as being in danger. This is the reason for the moves by Australian Aborigines and Native Americans to repatriate the bones of their ancestors, which are scattered in museums all over the world. The right to give them a proper burial would seem legitimate with regard to at least two considerations. First of all, the very long separation of these indigenous populations from the rest of the world justifies their perception that they own those remains. But there is also another even more important motive. The right to the repatriation of the ancestor's bones is a political issue based on respect for human rights and on the claim of equal dignity towards one's own cultural heritage (Tuniz et al. 2009).

If this patrimony is not symbolized by some tangible and long-lasting construction (as it happens in settled societies), and the oral tradition is not strong enough to maintain and transmit it, then the repatriation of the bones of the ancestors can constitute an important element of imprinting of the culture. And it is no coincidence that, after the discovery of Mungo Lady, a centre was erected near her burial site, which is open to the public so that Aboriginal people can pay tribute to the remains of their 50,000-year-old ancestor. In doing so, they also affirm their cultural identity.

A second reason to honour the dead who are not directly related to us goes to their status as leaders. We have already covered the first elaborate Gravettian tombs in Eurasia. Other rather complex burials continued in Palaeolithic cultures up until the dolmen tombs of the first urban settlements of the Neolithic, and to the pyramidal tombs of various civilizations of most recent history in Egypt, in central and South America and in Asia. Now we are dealing with very complex constructions which, through the cult of the dead, also speak of the societies they represented. The tomb of a chief, being a king, a priest or both, is also a symbol of social values. When societies are subdivided into classes, the shape of pyramids reflect both the division of labour necessary to build the structures, and the institutional organization of those hierarchical societies.

A third reason concerns the cult of special men and women to whom people turn when in need of comfort. In the catholic religion, the cult of the relics of saints and popes is widespread. In encouraging such habits, the church deepens its social roots and provides a service to alleviate the suffering of the believers. It tries to inspires people with good examples and role models. It also provides a sense of protection from misfortune and psychological help when fearing the future.

Finally, scientists form a fourth group who are interested in ancient human remains. They see bones and teeth as a precious biological archive of our past. There is a general scientific interest in being able to access them. Indeed, everything in this book is owed to the remains of our ancestors. Once, it was thought that the discoverers of a fossil were also their owners. The claim of repatriation of these finds, and the scientific interest in their availability, have both challenged this exclusive right. Knowing and appreciating the motivations of each group can perhaps help us solve some sources of conflict.

8.5 Bones, Tombs and Relics

Now the time is ripe for a brief digression into a very specialized field investigating the functioning of our brain. And not just the brain that we can observe today but also the brain we can reconstruct from the human skulls of the past. This will pave the way for the discussions that will keep us busy in the end.

Chapter 9
Brain and Mind

Many of us are convinced that no one can read our minds and that, if we are clever enough, we can hide our thoughts. Sometimes, our body language betrays us with small signs (pupil dilation, sweating, muscular tension, quickened heartbeat) that a careful observer could decipher. Lie detectors, or "truth machines", are very good at this. With a thermograph, for instance, someone could detect our lies by measuring the temperature of our noses (Salazar-Lópezf et al. 2012; Gołaszewski et al. 2015). But we usually think we're getting away with hiding our thoughts.

Today, we can analyse the functioning of the brain in a very detailed way, at least in living beings. But can we get inside the heads of our distant ancestors? Incredible as it seems, we can read their brains, and even their minds. What is the difference between these two concepts? Using the language of computer science and ignoring the centuries-long debate on this subject, the brain is often considered the hardware, and the mind, the software, of a data processing system. The mind would be an expression of the functioning of the brain. This current of thought is the frame of reference of the two major and much-debated projects to map our brain: namely, the Human Brain Project, funded by the European Community, and the Brain Initiative, supported by the U.S. Government. According to this approach, human cognition would depend mainly on the cerebral processes. With the mission of mapping the brain quite complex already, the roles played by body, culture and objects are disregarded.

To better understand cognition, some researchers prefer to study the so-called "extended mind" (Clark and Chalmers 1998) and consider tools and objects with cognitive processes, as well as the social environment, which all play a key role in enhancing our cognitive capacities. Those capacities can be seen as depending on an integrated system consisting of brain, culture and environment. In particular, brain and culture are connected by our body; culture and environment are connected by all the objects we use (Bruner and Iriki 2016). Note that these objects don't need to be "intelligent", that is, capable of interacting with us. It is sufficient that they are an essential part of our ability to interact with the environment.

Without fear of taking up the challenge of the "extended mind", the latest models proposed by the cognitive sciences[1] suggest that the brain modules responsible for our higher capacities, broadly defined as reasoning and abstraction, are not linked only to other brain components, such as the sense-motor system,[2] but also to the body, along with the natural and social environments. The new somatic extensions of our central nervous system have been recently confirmed by the discovery of direct connections between our brain and our gut (Hoffmann et al. 2018a, b). Other connections were mentioned some time ago and have been revamped in recent days.

According to the "cortical homunculus" model (Penfield and Boldrey 1937), the brain "represents" the different parts of the body and its movements in the environment—with a key role for hand touch—and considers instruments an extension of the body. Hence, if we look at the past, our lithic tools and ornaments would not only be something we think *about*; they would also be something we think *with* (Stout and Hecht 2015). Body painting, ornaments, garments and tools could then be interpreted as a transformation and extension of both body and mind. As one scholar put it: "It is no more possible to understand human cognition without reference to this fact than it is to explain the adaptations of beavers without reference to dams" (Stout and Hecht 2015). Hence our culture is not only assisted by the tools we use, as happens in some animals. It becomes dependent on them and is part of our thinking.

Another good analogy for the role played by tools and culture in relation with the environment is the spider web (Bruner et al. 2018). It seems that by considering the web threads and its configurations as an extension of the spider's central nervous system, it is possible to explain some characters of spider cognition, like route planning, for example. Spiders, along with other small animals, might have solved the problems of their miniaturization (which translates into an inappropriate brain-body scaling) by outsourcing their cognitive capabilities to peripheral parts of their body or to external components of their own making (webs, burrows, nests, artefacts and dams). In the latter case, adaptation processes would also fine-tune the properties and use of these extra-somatic elements.

In the case of humans, the connection between body and objects we use becomes part of our evolutionary mechanisms. It is transmitted between generations and influences the processes of reproduction and genetic selection. Through technology, we delegate an increasing number of tasks to outside of our central nervous system. By adopting the "extended mind" perspective, it is somewhat misleading to consider the brain as a sort of hardware and the mind as a sort of software. The mind processes interactions between brain, body and environment, which includes other human beings. The extended mind is a network that goes far beyond our skull. This description recalls the modern concepts used by physics to describe reality. As Carlo Rovelli (2018) would say, reality is made of interactions: no things "are", things "happen". The same applies to the way we think.

---

[1] For a broad discussion of these models, see Coolidge et al. (2015) and Stout and Hecht (2015).
[2] This system deals with perception, action and proprioception (the control of the position and of the movement of the body).

However, it remains difficult to apply the scientific method to demonstrate, in detail, how cognition extends beyond the neurons of our brain and how body participates in cognitive processes. Surely emotions and feelings play a very important role (Damasio 2018), particularly to promote the circuits of pleasure and reward through hormones and neurotransmitters. Probably they were also crucial in the self-domestication process, in analogy with what happened to some ancestors of our dogs and bonobos. Our mind is now being extended to include the artificial intelligence we are creating. According to Harari (2016), we could turn into gods. However, a change of perspective might be useful. After trying so hard to determine where in our body resides our mind—roaming among neurons, hormones and even intestinal bacteria—perhaps we should consider that our body is part of our mind.

We will see that the objects we use—according to the "extended mind" perspective—will be able to nurture, but also curb, our mental abilities. In particular, culture, environment, and related tools can induce inheritable modifications even if they do not directly affect our genome. This can happen through the epigenetic mechanisms, the processes that produce new anatomical or behavioural characters by the expression of our genes without involving genetic mutations.

9.1 Brain and Mind in Deep Time

The new methods of neuroscience mainly concern the mental state of current humans. However, they are also applicable, to some extent, to the study of our ancestors' brains. This might seem surprising. How do we obtain structural and functional information on the brain? It is one of the most perishable organs and is certainly not preserved in the geological record. In fact, however, there is a way.

Information can be gathered about an important part of the brain—the surface of the cerebral cortex. This is the rough outer part in contact with the skull. Its thickness changes throughout life, and it plays a central role in the cognitive functions governing self-awareness, language and memory. The external structure of the cerebral cortex is obtained by reconstructing the internal surface of the skull with X-ray micro-tomography.

We have already talked about these techniques in reference to the analysis of hominin teeth. If we now apply them to the fossil skulls, they can give us more information, and, above all, do it in a non-destructive way. In the past, the risk of damaging these precious finds greatly restrained the studies due to the methods available at the time. Today, the problem has been solved. In fact, it is possible to generate 3D images at very high resolution on which one can work in a virtual way.[3] By creating appropriate sections, one can find hidden details and make

---

[3]In addition, since the skulls are often discovered with many fractures and deformations, by using X-rays micro tomography, their original form can be rebuilt by virtually repositioning the fragments (Di Vincenzo et al. 2017).

quantitative comparisons between different brains. Therefore, even if the brain is lost, we have a detailed virtual copy of the surface of the cerebral cortex. We can infer the borders of the cortical regions from cerebral sulci, gyri and other landmarks.[4] And since we know the functions of the different parts of the brain, the internal details of the skull structure allow us to glean information about what the owner of the skull could mentally process and how (Bruner et al. 2018).

Obviously, the internal details of the brain in the deep past cannot be explored directly as they don't fossilize. However, it has been recently shown that this exploration can be performed indirectly by combining fossil skull analyses, paleogenomics and imaging of brains from extant humans. Genes involved in neurogenesis and myelination have been associated with the different shapes of modern human and Neanderthal brains (Gunz et al. 2019).

Even though many brain functions can be associated with specific areas of the brain, we know that connections are also very important. Unfortunately, we cannot get information on these connections without a functioning brain we can observe. New methodologies to study brain evolution are being explored in experimental "neuro-archaeology". It is possible to see, for example, the neural circuits that are activated when one of our contemporaries carves a lithic instrument with the same techniques used in the deep past. In moving to the processing of more complex objects, from Olduvaian to Acheulean instruments, an ever-increasing number of brain areas is activated. And since the neural networks stimulated during the processing of the stone coincided with those connected with the development of language, it was immediately thought that the Acheulean technological revolution was associated with the emergence of spoken language (Stout et al. 2008). But now it seems that it was not exactly like that. In more recent experiments, it has been discovered that the neural networks involved in Acheulean toolmaking are the same ones activated when we play a musical instrument[5] (Putt et al. 2017).

The evolution of the human brain has been debated for more than a century. If we integrate modern methodologies from various disciplines, we can open a new chapter of very promising studies. For example, we can line up the data on the external structure of the brains of the various human species against the respective genetic structures responsible for the various cognitive and behavioural functions. We can then line both of these up against the available archaeological evidence. This could include the level of complexity of the stone industry, the use of specific materials, personal ornaments, wall and portable art, musical instruments, the organization of areas dedicated to social life and the sophistication of burials. What can this connection and data correlation tell us?

---

[4]Or at least those that are imprinted on the inner surface of the skull through the dura mater, the most external part of the meninges, made of the membranes that surround the brain.

[5]The method used in these last experiments is called near infrared functional spectroscopy, a non-invasive technique that provides images on the functions of the human brain. It is based on the measurement of small changes in optical parameters when the brain areas of interest are activated by specific stimuli. In essence, it monitors local variations of certain forms of haemoglobin, induced by the cerebral activity.

Perhaps through the triangulation of the data, one could find out whether the evolution of the modern Sapiens mind occurred after sudden genetic mutations that introduced new structures or brain connections, generating new social and communicative behaviour. Alternatively, such a process could have been promoted by a combination of biological, cultural, demographic and economic mechanisms. In the latter case "circumstances" would have fuelled a development trajectory that had been latent and silent for many millennia. For a long time, these two interpretations were opposed to each other, stimulating a heated debate, but they might need to be combined. Various feedback mechanisms could have come into operation, some of a genetic and psychological type, others of a social and economic nature. They would have triggered virtuous cycles so as to lead to the evolution of the Sapiens mind of our day.

The new methods for studying the brain can be extended to investigations of pre-human hominins. The synchrotron imaging of the skull of *Australopithecus sediba* suggests that its brain, about as big as those of the present-day apes, had slightly asymmetrical frontal lobes, like those of humans (Carlson et al. 2011). This suggests that in the evolutionary line that preceded *Homo*, before the expansion of the brain, a neural reorganization was already under way. This reorganisation possibly involved an increase in the connections in the region that is associated with language and social behaviour in humans. Therefore, the separation of humans and previous hominins, who perhaps already knew how to make stone tools and had some cognitive abilities, becomes increasingly uncertain.

9.2 Thinking of Thinking

Some associate the dimensions of the frontal and temporal lobes, crucial for cognitive functions, with the dimensions of social groups and with "intentionality", defined as the ability to have convictions about oneself and others within a certain social context. Closely linked to the capacity for imagination, intentionality is often associated with terms such as, "I believe", "I think", "I wish" and "I suppose", and involves what people believe, think, wish or suppose other individuals think about themselves and others. It also applies to the psychological development of children, who possess a "theory of mind".

This ability, which we share with other primates, could be linked to mirror neurons (Rizzolatti and Craighero 2004; Rizzolatti and Sinigaglia 2016). These are particular neurons that come into operation both when individuals perform an action, and when they observe another person performing an action with the same goal. The process they activate enables them to perceive someone else's feelings as if they were their own, and makes people identify with what they observe. And this happens even when one is aware that it is a fiction, such as when one cries while watching a movie. However, the ability to understand and identify with the mental states of others does not depend only on mirror neurons but also, apparently, on networks of neurons that involve other brain areas (Mar 2011).

On the other hand, researchers have observed a linear correlation between the volume of the primate brain and the size of their social group (Dunbar 1995). This correlation has also been extended to the hominins (Gowlett et al. 2012; Dunbar et al. 2014). If we now combine this theory of the social brain and the theory of mind mentioned above and compare it with the psychology of present-day Sapiens in their early years of cognitive development, we can try a brief summary of what some paleo-sociologists have proposed (Gamble et al. 2014).

At the first level of intentionality, typical of a macaque and a Sapiens child of less than 3 years, there is only a certain self-consciousness. At the second level, typical of an australopithecine, a Sapiens child from 3 to 6 years and maybe even chimpanzees, one begins to discern what someone else thinks. At the third level, typical of *Homo erectus* and a Sapiens child of 6 to 8 years, one makes assumptions about what others think of him or others. At the fourth level of intentionality, typical *of Homo heidelbergensis*, and a Sapiens from 8 to 11 years, we can consider what we think of what others think about others. In normal adult Sapiens, this capacity reaches a limit at around fifth-order intentionality, with only a few individuals able to perform successfully at higher orders. It is at these last levels of intentionality that we see the onset of collective symbolic constructions. It is argued that this was the hallowed ground of the emergence of structured ideologies and religions.

For example, structured religions belong (at least) to the fifth level of intentionality, if some individuals want to convince others that by turning to a higher entity they can induce it to operate in the sense they desire, as suggested by their prayers. If we are dealing with religions that are claimed to be "revealed" by someone from above, we climb on another level. Proceeding on this path, in trying to convince you, the readers, that religions operate at the previous level, we, the authors, in turn raise a level of intentionality (reaching the seventh level). The reader could amuse himself by checking how many levels of intentionality he can manage, in a conversation, talking to his friends about what we say, and expressing an opinion about it. This reasoning could also extend to the case of ideologies, or even to advertising, or whatever is able to condition our behaviour. Note that scientific thought also operates at very high levels of intentionality, with the difference, with respect to religious thought, of always wanting to verify everything the human mind may have conceived.

9.3 Modelling and Imagining

The reader might think that we have opened up too many windows in our reasoning, and that it is now difficult to manage simultaneously all the issues we raised and present a coherent overview. After talking about how complex our brain networks are, we mentioned the need to interpret the brain networks of others (their thinking) and the networks of these networks (our relative thinking). All becomes even more complicated if we consider that, as we explained earlier, individual minds include not only the nervous central systems, but also bodies, instruments and the

9.3 Modelling and Imagining

nearby environment. Individual minds themselves are part of an extended social mind.

Fortunately, when matters become too difficult to handle, we know how to create models in which we place almost all the variables that "do not count" among the external conditions and we focus only on those we want to consider. This is what physicists do, for example, when they model the atomic nucleus, even if they do not know how to describe exactly its behaviour, and all the forces and particles involved.[6] In the case of our cognition, modelling is made more difficult because we lack clear ideas about the identity of the main "particles" or even the "forces" at stake. But it is simplified by considering that by generating a model, we do something very similar to what our brain does all the time. In fact, the brain selects, stores and processes, from a temporal perspective, the masses of information that our senses provide.[7] In the course of evolution, the senses have in turn been activated and sharpened or weakened according to the need for survival.

In our brain, a very powerful selective mechanism is put in place, to collect and use information, one that isolates "noise" and makes us focus on relevant information. This mechanism was brought to light by the Working Memory model (Baddeley and Hitch 1974).

This model will turn out to be very handy during our final discussion about the possible roles of artificial intelligence along our evolutionary path. For now, it suffices to say that it is based on the idea that some information is retained only for the time needed to perform some particular function. According to this view, despite the large amount of information available and the complexity of its interpretation, we can still count on a coordination centre for information and action planning. The Central Executive Module in the Working Memory model is based on selective attention. It suppresses irrelevant information and coordinates the relevant material. This module also organizes the perspective memory, assigning values related to the course of time to perform a certain number of operations in sequence. In short, even if information is massive, we have organizational models that allow us to manage it at an acceptable level of complexity.

The Central Executive module assists, and is assisted by, three other modules:

- a Visual and Spatial Notebook that maintains and processes information, generating mental images in 3D;
- a module called Buffer Memory, which retains temporarily multi-dimensional integrated information (such as scenes and episodes) taken from working memory and long-term memory; and

---

[6]The phrase written on the blackboard of the famous physicist Richard Feynman at the time of his death, in 1988, seems to have been, "What I cannot create, I do not understand" (http://archives-dc.library.caltech.edu/islandora/object/ct1% 3A483).

[7]We think we have five senses (and we call intuition our sixth sense) but there are many more, such as the feeling of pain, the perception of temperature (thermoperception), the sense of equilibrium, the perception of where the different parts of our body are in space (prioperception), just to name the most important.

- a memory of sounds able to retain and process both phonetic information (the production and perception of the sounds themselves) and phonological information (related to the meaning of these sounds).

In practice, just as researchers who want to interpret the functioning of the brain manage to model some of its functions, so does the brain when it models the surrounding reality by selecting the sensory stimuli coming from it. Focusing on some of the stimuli, it inhibits the perception of those that are out of focus. This is what happens when we watch a magician and do not understand where an object has disappeared to or how it reappears. Leading us to pay attention to a particular hand or movement, this skilled entertainer makes us focus on the wrong information. Fortunately, there are few magicians around. We are in charge. Or at least, we think we are.

At some point, modern humans started representing the natural reality that they had been modelling internally through their extended cognitive system, making the relevant images available to others. For example, using ochre pigments, they represented a lion that lives in a three-dimensional world and turned it into a two-dimensional drawing on the wall of a cave. Or they transformed an animated woman, with all her charm and beauty, into an ivory statuette with amazing sexual features. They represented their mood in a sequence of sounds (in music) or body movements (in dance). They represented themselves painting and adorning their bodies. They told tales of events and imagined worlds. When did these changes in the working memory turn us into modern Sapiens? Some suppose that the Central Executive Module changed between 60,000 and 130,000 years ago, following a genetic mutation that involved brain connections (Coolidge and Wynn 2005). Both the timing and the explanation are controversial (d'Errico and Stringer 2011).

The oldest evidence of the emergence of anatomical features compatible with a reorganization of the different modules involved in the working memory is the evolution of a globular skull in Sapiens, between 300,000 and 100,000 years ago. This characteristic is attributed to the expansion of the part of the parietal lobes in which the intraparietal sulcus and the precuneus are located. It seems that the development of these two areas is also responsible for the reorganization of connections between different brain networks (Bruner and Iriki 2016; Bruner et al. 2017). It is also attributed to the emergence of new skills, such as knowing how to count, developing a perspective memory, and generating new linguistic functions. Furthermore, the expansion of the parietal lobes is connected to the ability to imagine and understand symbols. Finally, it has been noted that a more globular skull could have provided the context for complex-language evolution, a sort of anatomical condition that, by implementing communication skills and optimising language transmission, might have coincided with the process of human self-domestication (Benítez-Burraco et al. 2018).

The upgrading of the Central Executive Module described by the Working Memory model can be called into question by the archaeological remains at Pinnacle Point, in South Africa, dating back 71,000 years. In this case, the production of flint blades required many phases of consecutive processing. It could therefore be completed only through sequential memory. The inhibitory function of the Central Executive Module seems to have been important instead to curbing violent behaviour in interactions with others and to promoting social cooperation.

Other primates have a certain short-term memory, and this also applies to many of our direct and indirect ancestors. But only in Sapiens did important characteristics appear at some point to put us on a particular evolutionary pathway associated with our demographic growth and our communicative development. It had to do with the urge to represent, or "perceive" reality through a mental translation of it. Landmark work has been performed, in recent years, to shed light on this impulse. Imagination seems to be key in this respect. We now know that we share with some other animals a number of capacities regarding communication skills, memory, a certain degree of reasoning, empathy and tradition. To model a distinctive human mind, we need more.

It has been suggested that what is uniquely human amounts to the capacity to perform mental time travel, read other people's mind, behave according to moral principles and provide abstract explanations to observed facts. Though the first three traits seem to be present in an embryonic form in some animals, full development is said to be possible in relation to the setup of a particular brain procedure. Basically, this procedure would consist in the construction of a "nested scenario", a sort of inner mental theatre in which we can envisage a number of situations and anticipate their possible outcomes. By visualizing events that are not yet occurring, we set up a sort of stage, where actors are joined in a presumed narrative. This imaginary play takes place under the supervision of a director (who evaluates and creates the scenes) and the control of an executive producer (who makes the final decision about what to pursue) (Suddendorf 2013; Suddendorf et al. 2018).

Still, the future remains uncertain and therefore a second feature emerges: a wish to exchange our thoughts with others. As nobody owns a crystal bowl, to further reduce uncertainty, we ask those who have already experienced some situations to tell us what may lie ahead of us. In doing so we move from an individual intentionality to a shared intentionality. We can thus reach the best of both worlds. We can nest in a specialist niche of personalised narratives as individuals and, at the same time, enjoy a generalist niche of cultural narratives as a species. This is exactly what risk-averse investors do in finance when they diversify their portfolio and hedge against unforeseeable situations. The development of what are considered the distinctively human mental capacities may have followed the same logic: cherry-picking from a vast smorgasbord and asking our fellow partygoers if that unknown delicacy tastes good or bad.

Note that this model focuses on the development of such capacities according to age. In fact, both functions, embedded in our brain, develop at different rates during childhood and map into the construction of a working memory, recursive thought and executive function all the way into adulthood. In other words, younger people are still unfit to imagine complex scenarios and less inclined to listen to the experienced, whereas older people are more disposed to do both. It is therefore tempting to extend this reasoning to the domestication hypothesis that might have affected the evolution of modern humans and imagine what would happen—

according to this model—if people keep their juvenile traits into adulthood. Is there a risk that our shared intentionality will diminish? Shall we perhaps try to perform tasks and achieve goals disregarding the multiple possible outcomes of our actions? Shall we underestimate the experiences of others and take their advice with contempt?

If this were the case, we might end up taking unnecessary risks and acting incautiously. In a heavily armed world this prospect is certainly not reassuring. Alternatively we could fall—with pleasure—in the arms of those who will save us a lot of troubles and take full responsibility for our actions. Exactly like children do. We will see that artificial intelligence is the most likely candidate to perform such a role in the years to come. Indeed, it already does, to some extent.

Chapter 10
Imaginary Worlds

When we encounter a work of art, be it a painting, statue or musical composition, we tend to focus on the work itself, trying to understand the creator's message and whether it reaches our heart (actually, we should say our brain and include the circuits of pleasure and reward triggered by hormones and neurotransmitters). We seldom ask ourselves why, among all animal species, we take the trouble to represent what we see, hear, and feel, and share it with others for purposes apparently unrelated to survival.

We know that bees transfer, through dance, important logistical information to the whole hive. And that many other species communicate. But that information centres on food, the willingness to mate or swarm and the presence of danger. Some animals can also provide information on their fitness. The dances of some insects (such as fruit flies) or the songs of many birds can act as a sexual call between males and females. Bonobos have been trained to communicate using visual symbols. Some parrots can name tens of objects and transmit their wishes. But it is a bit far-fetched to imagine that they also communicate what they think, in the absence of evidence. Still, as the saying goes, this is not "evidence of absence".

In contrast, modern humans are not satisfied with their perception of things. They want to share it. By inventing a spoken language, they first represented an object through a sound and then connected these sounds in a grammar and syntax that form a structured complex language. By writing, they replaced the sound with a fairly permanent sign. Communication is based on signs and symbols to which we attribute a shared significance. A symbol refers, in particular, to a reality that must be recomposed in the eyes of the observer. In Sapiens, the emergence of these new abilities can be traced back to around 100,000 years ago. Since then, symbolic thought had increased dramatically in all possible expressions and they have gone from a few thousand individuals (the small group that we can genetically identify as our direct ancestors) to more than seven billion people, a number that continues to increase. Did any new trait emerge, during that period, which could be responsible for such a result? In what follows we argue that, besides complex language, symbolic thought and culture accumulation, another key trait was the ability to create and exploit emotions.

The mechanisms responsible for the development of a complex language are still debated. Initially, language structure was treated as a genetically encoded biological trait, and said to emerge as a result of a complex adaptation for communication (Pinker and Jackendoff 2005). But there is neither a single switch, nor a single gene, accountable for its appearance. A whole network of genes could be involved in syntactic processing and speaking proficiency (Burenkova and Fisher 2019). However, many aspects of language structure seem to appear as adaptation of language to constraints imposed by the way it is transmitted (Kirby 2017). Extensive work has probed into a range of linguistic phenomena. Much less has looked into the implications of these findings for the evolution of language. Whereas biological evolution allows individuals to convey ideas according to the cognitive and physical abilities they receive from their parents, emerging language allows other individuals to understand and retransmit them according to their own capacities. This emerging platform of collective learning can in turn have feedback effects on some biological characteristics. Thus, language develops through a virtuous cycle, interweaving cultural and biological evolution, as well as individual and social learning (Arbib 2012).

In particular, if one considers language as emerging from cultural transmission, two traits enable cultural evolution to create linguistic structures and act as precursors for language to develop in the first place. One is learning. Cultural evolution occurs if something is learned and transmitted between generations. Another one is the ability to discern an intent. Learning needs to be guided by the capacity to recognise when a certain signal or action is aimed at conveying a message. Strikingly, both precursors are associated with domestication and have been observed in other extant animal species. In particular, the first trait was seen emerging from the domestication of the Bengalese finch, who can only perform structured singing if taught to do so. The second trait was observed in the domesticated dog, who can infer intentionality from a number of vocal and body signs. Only modern humans, however, are endowed with both traits. These findings suggest a key role of self-domestication in the evolution of human language and seem to confirm that the evolution of language structure was probably born at the onset of a process of self-domestication (Thomas and Kirby 2018).

Recently, the passage of language down to a generation of learners has been studied in a novel way by creating digital models of speakers who had to learn language from other digital agents and transmit it to other agents. The results are striking. When they pass it on, agents tend to produce a more "structured" language than the one they have received as an input. In other words, when a random string of words appears to be slightly ordered, the learner "imagines" a structure and then reproduces more structured words in its output. A cumulative accretion of linguistic structure has appeared (Kirby 2017). The reproduction of this experiment in humans and animals is now under way and promises to deliver thrilling implications.

So far, we have not discussed the content of what we want to communicate via a complex language. Surely it must have been about a general improvement of our condition for survival. Yet, at a certain point, we started to pay an extraordinary amount of attention to our looks, adorning our bodies in all possible ways. This care spread to the appearance of the environment we lived in and included the objects we used. Through recognizable signs, we were adding value to ourselves and our surroundings. We began producing shell necklaces and bracelets worked and intertwined in different styles. We did drawings, tattoos and scarification on our bodies. We carved figurines of men, women, animals and inanimate objects. We drew shapes and figures on every available surface. We left signs of our passage, signing the cave walls with the stencils of our hands. Besides inventing a structured language to communicate facts and ideas, we also wanted to set up a body language to communicate who we are, a space language to communicate how we live, a societal language to tell which family, group or tribe we belong to, an artistic language to reach out and show how good we are in interpreting the world around us. In a nutshell, we also wanted to speak languages that could arouse emotion and would click according to the capacity of the beholder.

Most of this behaviour is documented by archaeologists, with complementary information from ethnographic studies. Some behavioural traits, more volatile and ephemeral, can be reconstructed only with careful anthropological inference when, for example, we connected through stories generating myths and legends. Or when we communicated our feelings through a choreography, a sequence of sounds or the composition of a song. Or when we celebrated collective events that united us. In all of these cases, we generated messages, objects, situations and actions that were beneficial and exciting. For those who maintain that "first comes duty, then pleasure", it is perhaps ironic that our chances of survival had increased by activating, among other things, our neurotransmitters of pleasure.

Returning to the genesis of this behaviour, based on a new ability to generate and exploit emotions, the possibility of spreading and articulating symbolic thought increases if the barriers between individuals are reduced in space and time. That is, if someone who represents an idea does not necessarily have to meet physically, or be contemporary, to those who recognize and interpret it. This is why the invention of writing is so important for the spread and accumulation of knowledge. Since then, our communicative interactions have dramatically increased, and recently boomed, via intelligent technologies and digital networks.

On the other hand, we have seen that, by using symbols, we can also promote important feedback effects on the functioning of the brain, which becomes more receptive to the ideas of others, with undoubted relational advantages. In forming wider and more interactive societies, our new cognitive capacities activate at least two social mechanisms: a multiplier effect for the spread of ideas (which depends on shared languages and the frequency and number of social interactions) and an acceleration effect in the formation of these ideas (which depends on the feedback mechanisms of shared ideas on the brain). One would think that we have set out on a path leading to endless progress. But we will see that this path can be very bumpy.

Now, we ask two important questions: How is culture accumulated and fixed in a population? And what do we mean by culture? Among all definitions, we refer to culture when a certain behaviour becomes stable and is rooted in a group of individuals, passing through generations. It is partly stable and partly evolving when new ideas are established. These ideas would work similarly to genetic mutations, but genetic mutations are random, whereas new ideas often arise for precise purposes and are adopted only if they enhance survival chances. With this definition of culture, certain behaviour, such as the control of fire or the production of stone tools, can take root and be passed down through generations for their simple practical utility. We could then find cultures before the formation of symbolic thought, cultures that did not change because they were perfectly able to guarantee survival in the absence of major environmental challenges.

There would therefore be a fundamental difference between symbolic thought and culture in the sense just mentioned. While symbolic thought, once born, emerges continuously in the course of the various processes of abstraction, culture is defined in a time and in a circumscribed space. In other words, symbolic thinking is a process, culture is often a product of this process. The former needs to manifest and spread, the latter to take root. The first consists of a flow of ideas that multiply, the second in a stock of knowledge to be shared.

Accordingly, many cultures have emerged in all of the Eurasian areas populated by Sapiens during the late Pleistocene.[1] But it does not follow that other cultures (in Africa, Asia and Oceania) are to be compared with the European experience to create a hierarchy between lower and higher cultures (Mcbrearty and Brook 2000). Nor does our definition imply that there are no cultures without symbolic thought. The Acheulean culture, for example, consisted of information and technology replicated with minimal improvements for about 1 million years.[2]

On the other hand, according to our definition, the *Homo* species are not alone in having a culture. Whales and dolphins lead very rich social lives in which they are called by name and "speak" to each other. Researchers think that they engage in some sort of "gossip" and therefore have advanced cognitive skills of a pro-social kind (Fox et al. 2017). Studies are flourishing in this field. Scientists were even able to assess the degree of attention of the Capuchin monkeys as they learned traditions (Fragaszy et al. 2017). Therefore, symbolic thought acts as a catalyst for cultures to develop and evolve.

The archaeological evidence suggests that symbolic thought probably emerged only in *Homo sapiens*, between 100,000 and 50,000 years ago, in South Africa, as far as we know. As long as it is transmitted and replicated, it leads to ever-increasing cultural complexity, documented by more and more sophisticated ornaments and stone tools. According to some scholars, this complexity is related to population

---

[1] Rich archaeological documentation actually testifies to its presence in Europe starting around 45,000 years ago. In a timescale that encompasses the degree of technological and artistic complexity, we speak of Châtelperronian, Uluzzian, Aurignacian, Gravettian, Solutrean and Magdalenian.

[2] It is widely accepted that the producers of this technology meant to give the artefacts an almond shape, even if they probably did not have a detailed image of the shape at the beginning of their work. Yet, they had to use more complex thought than that associated with the production of more archaic stone tools (Coolidge et al. 2015).

density (Powell et al. 2009). This means that endless progress is not guaranteed. It might happen that, as the population decreases or spreads, a society loses the collective knowledge of the past. This is what probably happened between 70,000 and 40,000 years ago, when the previous cultural brilliance disappeared in Africa and new cultures flourished in Eurasia. The same holds after the isolation of Tasmania from the Australian continent, due to rising sea level 12,000 years ago. In this case the complexity of the culture of the new island, Tasmania, remained frozen compared with that formed when the great continent of Sahul existed and then gradually decreased over the next millennia (Flannery 1994).[3]

In the opposite case, when the population reaches a certain density, symbolic thought develops and extends into an increasingly complex series of representations. The social expansion of symbolic thought seems to be limitless. After having admired all its outcomes, at a certain point all these representations seem exaggerated, even paroxysmal.

10.1 New Realities

It would seem that, in our species, there has been an evolutionary advantage in expanding the dimensions of our groups and strengthening our social cognitive abilities, in a virtuous cycle. This feature also extends to the economic sphere, for it allows the social product to increase considerably through the division of labour. If this greater product is redistributed with equity, the well-being of the general population also increases. And the list of benefits to create larger communities could continue. The problem is how to manage the cohabitation and social interactions of groups formed by many people. We have seen that understanding what others think is very useful. But is it enough, in all the situations of our daily lives? And how did we manage to organize communities formed by millions, and today, by billions of people?

The answer is very simple: by creating imaginary realities. If these realities are made of stories that everyone is supposed to believe as an act of faith, we speak of religions. If they are made of arguments about the best way to live in harmony and justice—or to dominate and suppress other populations—we speak of ideologies; if they comprise a set of rules that everyone must conform to, we witness the onset of states, nations, and institutions in general. Sometimes, stories and rules overlap and the adherence to a belief is associated with a certain behaviour. The keys that allow us to enter that imaginary world are made of symbols. But mind you, in contrast with Harari (2015), saying that these worlds are the result of our imagination does not imply that they are not real worlds, and therefore "fictional" (i.e. "not true"), just because we cannot filter them through our senses.

---

[3]The native Tasmanian people would then be decimated with the arrival of the Europeans.

By linking via languages, by setting roles for individuals and rules for the community, by identifying ourselves as members of that community—in a nutshell, by creating cultures—we form a social organism that is "real", and bears real consequence, as long as we share a common confidence in it. The same holds for science and any other product of our imagination. But whether an idea, or a set of ideas, is "true" or "false" is a different story, as this particular feature depends on the boundary conditions of each application of reasoning and the extent to which a certain statement can be falsified. Furthermore, abstract ideas can leave some members of a community better off, when it comes, for example, to the redistribution of the social product. Rather than throwing all ideas in the same basket, and submitting them to a reality test, it is perhaps useful to sort them out—at least some of them—and check for their purpose and validity.

Having said that, no wonder we are surrounded by so many symbolic images: deities; sacred buildings; holy rituals; colourful flags; icons to form national and local communities; work tools, such as the scythe and hammer, to suggest an ideal of equality; and heraldic crests to signify obedience to a hierarchy. Other institutions are more elusive and do not like to use explicit symbols, because they claim to exist in their own right. The Stock Exchanges are temples for those who believe in the market. And if they begin to doubt it, the market collapses. The market itself is an abstraction. It is just a rule of thumb that rests on the "ability to pay" of each individual, under the so-called laws of supply and demand. Indeed, there are other rules for sharing scarce resources. For example, by making long waiting lines for goods and services. That is what we do every day in a post office, and what people had to do in Moscow, a few decades ago, when the price of bread was kept low for social reasons. Other abstractions are more elusive and bear hidden consequences. Think of the invention of agents with legal definitions.

By forming companies with limited liability, very few managers pay for their mistakes or wrongdoing. In fact, they often leave the company with high compensation payouts. And the owners who appointed them limit their loss (at worst) to the value of their participation in that enterprise. Those who pay the higher price when decisions are ill-fated are the creditors and workers. If the creditors are banks that risk bankruptcy, their depositors might be involved as well. If banks are bailed out by a state, taxpayers will be the losers. In the end, two categories of people will be generated: those who make decisions and will pocket the relative gains if things go well, and those who are affected by these decisions and will carry the losses if things go wrong. By assigning symbolic values to some conventional object, we make sure that everyone exchanges it at that value, and so we transform it into money. We could even generate virtual wealth through electronically executed transactions without any exchange of goods or services as long as we believe that we can always draw on this wealth.

In politics, individuals can hide behind the institutions they represent and think about getting away with destabilizing an entire region by subjecting it to endless wars. Others can think they are heroes by performing terrorist attacks. The sufferers will be people who have not participated in any of these decisions and who do not have the slightest idea of the real reasons for their fate. So, symbols, and related institutions, can be interpreted both as a social glue and as a means of disguise from the real issues. But neither is fictional, just because it is born in our minds. Even if one cannot see, touch or smell a prime minister, for example, but only the person who is called a prime minister, once a community has embedded the basic principles that support the idea of a prime minister—e.g. democracy, the rule of law, the division of power, etc.—that figure immediately gains legitimacy and bears material consequences. Even Santa Claus "exists", as long as there are quivering expectations in children's minds and a load of presents under the Christmas tree. But there is no more Ceausescu—President of Romania—after the booing of his people, during the last speech he delivered in Palace Square (now Revolution Square) in Bucharest. That person, in flesh, was just someone to get rid of. And that institutional figure was something to be replaced as soon as possible. Still, the story about Santa Claus is fictional, and the story about Ceausescu is not.

We certainly derive many advantages from the pervasiveness of symbolic thought to the extent that it unites people. But we also see a progressive depersonalization of many actions, and therefore a lack of responsibility for their consequences. For Sapiens, and only for us, it seems that there are no limits on proposing possible worlds, all based on completely invented narratives. If we share them, these worlds lodge in our minds and become real. But only under certain conditions—that they are first represented with a symbolic image, then recognized and shared by a community, and finally, incorporated into the community's cultural heritage. In the end, we add a fifth dimension to our lives: one based on all the products, ideas and social relations that we can imagine and sustain.

This dimension is called culture, and in humans (only) is made both by material and immaterial products. Eventually, in association with our cognitive capacities, and considering all the feedback that we are trying to elucidate here, this fifth dimension allowed for an extraordinary empowerment of our species. The processes that favoured the onset and spread of symbolic thought in early modern humans has been quite controversial.

Some have argued that, in conditions of great environmental adversity, alongside the survival strategy based on natural selection, it was useful to strengthen another evolutionary tool: sexual selection.[4] This mechanism would have come into operation by activating and expanding features already present in our species and used widely for mating. The new neural networks could be combined with the ancient physiological traits generating emotions. The reproductive chances would increase for individuals endowed with greater "artistic" creativity: those who knew how to better represent themselves through ornaments, body paintings and objects capable of attracting the other sex. In particular, since body art seems to have preceded or

---

[4]Natural selection depends on the success of both genders in environmental adaptation; sexual selection depends on the relative success of same-gender individuals for the purpose of mating (Miller 2000).

coincided with the first artistic expressions,[5] one could also seduce with the help of the mind. Later these characteristics could extend to other fields and include music, dance and the development of new technologies to attract all kinds of people, of both genders.

A second explanation for the emergence of symbolic thought refers to our ability to generate a complex language suitable for relating to an ever-broader and more structured social organization. We have already touched on this argument. Here the reasoning goes as follows. In forming larger groups, beyond a certain threshold it would not do to rely on a communication system capable of describing only the external world (for example, a danger signal common to many animals). Neither would a language serve that allowed us to exchange information about ourselves and our acquaintances to build relationships based on trust (possible only within a small society). To draw together and organize many people, it would have been necessary to invent a language capable of transmitting an ideal construction. To adopt more socially oriented conduct, we might have resorted to cultural evolution, exploiting the tools developed through sexual selection. Besides competing within our own gender for the favours of the opposite sex, we might also compete with everybody else to seduce people in general—by creating emotions—and thus be popular, loved or respected.

Contrary to what was often argued during this debate, we think that explaining creative behaviour by (extended) sexual selection is not necessarily antithetical to a cultural evolution based on the affirmation of a complex language. Indeed, both explanations reinforce each other. As social numbers increase, acquiring a complex language helps to seduce with the mind. And seducing with the mind requires the adoption of a complex language. However, there is a substantial difference between human evolution through sexual selection and its transformation into the social evolution of modern humans. The former allows one to seduce many times (in a strict sense) to generate the greatest number of descendants. The latter allows one to seduce (in a broad sense) multitudes of individuals who endorse that symbolism and feel part of that "ideal society". Eventually, due to our endocrine system, individual pleasure is sublimated in collective pleasure derived from a sense of belonging to a great and magnificent social organism. To present and future generations, we not only transmit our genes but also our visions of the world. It is to that vision that each and every one is asked to conform, in a struggle to capture a suitable position in the social body.

---

[5]The use of ochre powder (probably sprinkled on the torso and hair) is tens of thousands of years old and precedes the appearance of rock paintings in Eurasia (Watts 2009).

10.2 Overcoming Perceptual Barriers

The importance of symbolic thought is not only about artistic, political or economic representations. It also concerns the progress of scientific knowledge. In allowing us to conceive worlds that we do not see, it frees us from the false information provided by the world we do see. This information, filtered by our limited and imperfect senses, is dazzling. Even scientists and philosophers are sometimes fooled by perceptions: It took thousands of years to understand that the Earth is not flat or that it revolves around the Sun and not the other way around. Our senses also prevent us from grasping reality outside a narrow range of colours, sounds, smells, tastes and tactile stimuli, to include only the main senses. Our sensations are then filtered by a mental reconstruction that approximates them on the basis of experience. Through modelling, the brain provides us with the perception of things.

Sometimes our perception of reality can be distorted. We are led to see things that are not there, or to be blind to things that are there. Both effects depend on incorrect brain decoding of input data, within our sensory signal band. We can also have hallucinations, or perceptions generated autonomously by our brains, independently of sensory stimuli. We know that they are associated with the shortening of the paracingulate sulcus in the prefrontal cortex (Garrison et al. 2015). But even in the absence of this pathology, sometimes hallucinations happen autonomously, such as when we are deprived of any sensation in a dark and soundproofed cell. We begin to see and feel things; our brain needs stimulation. If we do not provide any, it produces it itself.

Imagining worlds also means going beyond the set of sensations and perceptions created by our brains on the basis of the information gathered by our senses and its interpretation, correct or incorrect. We can even "see" the invisible components of matter, such as atoms and quarks. But first we had to imagine them. And to do that, we had to build on knowledge already acquired, review it and proceed to acquire new knowledge. Besides affecting sociality, symbolic thought helped us understand the natural world, changing the fate of our species and that of the entire planet. That was after we reached a critical mass of acquired knowledge and the intensity of communication. This holds both for the interaction among humans and for our communication with intelligent machines, which in turn communicate with each other.

Now, with the new information and communication technologies, we can connect the entire world population. It is therefore possible to exponentially accelerate our visions of the world. A new protagonist has entered the stage: artificial intelligence. For now, it is one of our allies. But it remains to be seen if this will always be the case.

10.3 Excess of Representation: The Economic Sphere

Unfortunately, there are some drawbacks to all of the above. Our artistic production dazzles us in its abundance and variety. How can we explain the hundreds of thousands of statues, paintings, and objects worked and decorated and left for us from the different cultures we know? Were they all necessary to symbolize the skills and ingenuity of a population so as to form a collective identity? If we limit ourselves to the world of art, these concerns might seem exaggerated; there is no limit to beauty! But what happens when these exaggerations extend to economic forms? Is there a limit to utility?

Think of when we increase our wealth to an extent that is largely superfluous to satisfying all of our possible needs, present and future. Is this an excess of generosity to our descendants? We doubt it. Certainly, it makes sense to want to die well-off, because we can count on the best care and assistance before departing forever. But what sense does it make to die immensely rich, since we can't take it with us? The only reason that comes to mind is that with money, besides acquiring material goods, one can rise in the social hierarchy, and once at the top, be "in power". But to what end, since these people are already rich? What is so fascinating about power? Is it perhaps a path to immortality, first in other people's mind and then, perhaps, in real life?

One is tempted to suggest that, beyond a certain threshold, we are dealing with a psychopathology. But if we accept that our species has taken an evolutionary trajectory that leads to a certain degree of self-domestication, this apparent absurdity is justified. The reason lies in the stratification of social classes, according to a pyramidal scheme. Those who try to enrich themselves far beyond their present and future needs do so because they are seduced by the consumption patterns of the more affluent classes which dominate them. At the same time, they want to seduce, with their riches, the subordinate classes, which they try to dominate. In doing so, they revere the "upper" classes by which they are fascinated and want to be revered by the "inferior" classes, which they want to fascinate. It is a hedonistic motivation, sometimes characterized by a good dose of narcissism, which has as its trigger what economists call the "demonstration effect" (Duesenberry 1951). In a bold simplification, no matter how good we may feel right now, we always want to feel "better" if someone else pretends to do so. And if we live in the illusion that material goods will do the job, there is no limit to the quest for an ever-increasing economic status. And to the power that goes with it.

The exaggerations in the economic sphere do not end here. We also exaggerate in attributing a value to our present wealth. Every day, we exchange financial products to which we have assigned an estimated value that is certainly a multiple of that of the real assets that they should represent. And in the past 30 years, we have circulated credits that represented (in excess) some kind of real wealth, but also credits that represented other credits, and this has happened many times, generating the so-called "financial bubbles". Those who hold these financial products believe that they own the property of some company, some assets or some receivables. And if they do not believe it, they think they can convince someone else to believe it.

As long as everyone lives under the illusion, no problem arises. But if one day this truth came to the fore, many traders would have to erase this virtual wealth from their wallets, since a good part of it is only in the minds of those who believe in it. In a totally interdependent system, such as the one we live in, this would not only be "their business". It would fall on all of us, generating "financial crises", with relative "recessions" (when the effect is contained) or with consequent "depressions" (when millions of families are on the streets). The foundations are laid for the next war.

In this regard, digital technology, which now guards and generates a good part of our virtual wealth, is often accused of amplifying imbalances, especially when "out-of-scale" variations occur in daily trading. But the real problem, which we often ignore, is that digital platforms are now the main safe of the virtual wealth of humanity, the management of which is increasingly entrusted to digital agents. We can already anticipate that these are very vulnerable and dangerous systems. Vulnerable because they are subject to penetration and manipulation by other faceless humans (some authorized to do so, others not) assisted by machines. Dangerous because those who control them do not hold financial power alone; they hold power tout court. Who would argue that our democratic systems, subject to the wish and fads of the multitudes, are still an effective antidote to this power?

We can therefore conclude that, even if one would limit himself to defending his current interests, and remain in the prevailing cultural and technological paradigm, it is necessary to contain our excess of representation. Unfortunately, we do not know how yet. And as long as we can manage the phenomenon, for example by deploying a vast array of weapons of mass distraction—and this is not a typing error—it's nice to continue playing in the big casino into which we have transformed the world economy. Or at least, that's what the players think.

10.4 Symmetric and Asymmetric Warfare

On a different level, we also exaggerate the potential exercise of force. We could destroy our planet, as we know it, many times over. Although states sometimes try to reach agreement on a reduction in the weapons of mass destruction, each contender has fire power which is a multiple of every possible annihilation of the adversary and his environment. Our global security does not depend on the relative might of adversaries but on how much each fears the possibility of retaliation to a first strike. To save us, for the moment, is the logic of deterrence, so we limit ourselves to fighting asymmetric wars. These wars—between states and non-state entities—emerge from many causes, and are fought mostly in disguise: hegemony over scarcer resources can hide under religious differences, control of illicit drug trafficking can pass as a fight for democracy (or autocracy). It is often argued that these wars cannot be won, as long as the underlying problems persist, unless one combatant loses interest in the fight.

Is there perhaps a hidden rationale behind this situation? Maybe there is. If we apply the domestication hypothesis to geopolitical matters, and consider ourselves as driven to climb an ideal pyramid of power among competing states or cultures, under the influence of a "demonstration effect", we can paraphrase our previous remarks, and argue that no matter how safe we may feel right now, we always want to feel "safer" if someone else pretends to feel so. And if we live in the illusion that piling up endless firepower will do the job, there is no limit to the quest for an ever-increasing armament. To flex our muscles, and flood with emotions a crowd of devotees, what's better than a mighty military parade?

From a slightly different viewpoint, it is as if on our evolutionary trail, we forgot to insert a stop command, and after having enjoyed the benefits of cooperation *with* competition in our evolutionary path, we found ourselves struggling with the need to invent new foundations for coexistence on a very large scale. If the excesses of representation in the world of art can end up harmlessly in the cellars of museums or the safes of millionaires, in the world of economics and in politics, these excesses could harbour a command of self-destruction. If what we have said has any substance, the symbolic thought that had favoured the formation of ever-bigger pyramidal societies is now faced with at least two bottlenecks, one of an economic nature, the other of a geopolitical nature.

On the one hand we see the stratification of wealth and the hoarding of the last non-renewable resources by the ruling classes of the richest countries and, within these, by an increasingly restricted group of people. Surely, the standard of living has improved in most "emerging" (now emerged) countries. However, on a global scale, there has been an impoverishment of past lower and middle classes and a widening of many areas of poverty (Stilwell 2019). Note that to feel "worse off", this impoverishment does not need to be absolute. If we are under the effect of a domestication syndrome, a relative impoverishment is enough to generate social unrest. For instability to arise, it is enough to feel excluded from an "equitable" distribution of the benefits that are commonly generated, according to our cultural scale of values.

The latest data show that one per cent of the world's population now holds the same amount of wealth as the remaining 99%.[6] With wealth unevenly distributed on a grand scale, it is difficult to manage conflicts (internal and external) by the old symbolic frames of reference. When gasoline is poured all over our past institutions, a little spark will suffice to burn down the house. To defuse tensions, we ought to invent new principles of social coexistence. And imagine new symbols beyond flags and deities. Old symbols are progressively losing ground—in our minds—in favour of the icons of our digital devices, which are simply associations of an image with a function. Unfortunately, icons will not suffice to substitute for lost symbols and might indeed produce the opposite effect: a disruption of social bonds, based on shared ideas.

---

[6]World Economic Forum Annual Meeting, 2018, www.weforum.org.

10.4 Symmetric and Asymmetric Warfare

On the other hand, we observe the formation of various hierarchical layers, in the military capabilities of the states, which might cooperate in some areas (e.g. by sharing some piece of intelligence against a common enemy) while confronting each other in other fields to boost their geopolitical supremacy. These complex and variable structures of interstate relationships are based on the institutional exercise of force. They are also traversed, now, by asymmetric wars, which are partly fought on conventional battlefields and partly "behind the lines", that is inside the enemy's territory. Guerrilla warfare and terrorist attacks, such as hijackings and suicide bombings, are notorious examples of asymmetric warfare. In these latter cases, their strength is wired in the weaknesses of the adversary: for example, in his obsession with security and control. It may therefore happen that the (hidden) strategic objective of a certain military action—for example the abolition of human rights that go against certain religious principles—does not depend on a terrorist attack in itself, but on the opponent's defensive reaction to it: a reduction of personal freedom. Like in a pool game, the best players are those who can play off the cushion. Too bad they are also playing, all too often, behind the curtains of deceptive institutions.

The refusal to assign an institutional image and name to the enemy sometimes forces contenders to use improper expressions to identify it, preventing us from recognizing the real adversaries and the real problems. So, somebody may fight "terrorism", which is simply a technique of war, or "religious fanaticism", which is simply a form of recruitment. And another one fights the "devil" and his impersonations. It has been a long time since enemies faced each other on a battlefield with their beautiful signs of recognition and their identities obvious, even if symbolic. This is when they fought for unequivocal reasons.

In today's total confusion, we must invent new principles of peaceful coexistence with related symbols. Obscured by the shadow of nameless hawks, the doves are no longer doing their job. To manage and prosper in a world populated by seven billion Sapiens on the eve of the depletion of the last non-renewable resources, we must change pace on several fronts. We have said that when certain ideas are rooted and handed down, a culture is generated. That culture can evolve if new ideas can penetrate a society. It is perhaps time to imagine a new Anthropocene: one in which we will introduce acceptable criteria of sustainability and restrain, as far as possible, our excesses of exploitation and destruction of human and natural resources.

Some say that a new "species" is already forming within our social body: a vanguard which occupies minority cultural niches, despite the fact that it is strengthening. It would be made of people who respected the lives of all living beings and who adhere to new moral principles. They would be the spearhead of a new society defined as post human (Caffo 2017). To assert itself, this new species would need new conditions that would push Sapiens to its limits, something that the advent of artificial intelligence could accelerate (Barrat 2013).

Chapter 11
Homo Oeconomicus

In this chapter, we will take a closer look at our economic behaviour. We will try to understand what it means to adopt such conduct and when we can trace the onset of this attitude. In order to do so, we need to get rid of the fog generated by an ideal construction of how we should behave in theory, instead of taking observable facts into account.

When we described the onset of symbolic thought, we stressed the need to represent something that we felt within ourselves. We referred mainly to the world of the arts, dismissing the craftsmen who imagine in objects new combinations of designs and materials, to make things both useful and beautiful. If we refer to the construction of something useful, we are entering into economic issues. We have made several incursions into this field on the previous pages. Now we ask: when was *Homo oeconomicus* born? This fictional character is central to the fate of our species and, perhaps, to the destiny of the entire planet. And what exactly do we mean by this term?

According to the most popular definition, a *Homo oeconomicus* is one who, in a situation of scarce resources, merely pursues self-interest, and, in doing so, benefits society as a whole. In a group of hunters and gatherers, however, the idea of an individual who behaves in this way makes no sense. Indeed, the opposite is true. If the results of one's efforts are uncertain, the survival of each member of a group depends upon the collective success of the communal hunting and collecting. In that society, nobody would dream of keeping for himself the abundant results of a day's work if the others returned empty-handed. Not every day is lucky. It is the collective success of the group that feeds each individual every day.

It follows that those populations, typical of our deep past, were not made up of men and women who were economically minded, and that this hypothetical individual was born, conceptually, only with the advent of agriculture and industry, when it was convenient to exchange goods and services and make the most out of this new way of dealing with each other. Indeed, the first metal coins seem to have appeared right then. It follows that *Homo oeconomicus* is no more than 10,000 years old. This widespread viewpoint, however, is at odds with recent evidence.

First of all, the basic definition of *Homo oeconomicus* as stated above is not satisfactory in the long run. On the one hand, it is too broad. Who does not try to pursue his own interest? And, on the other hand, it is too restrictive, for it usually refers to a person who chooses what to consume and what to produce given a certain amount of scarce resources. This *Homo oeconomicus*, whose behaviour is taught in every class of economics for beginners, can increase his well-being only by making a virtue of necessity, that is by being satisfied with what he can afford. Yet if we want to explain why this hypothetical individual behaves the way he does and insists on focusing on his own needs, we should refer to the role of psychology. For example, how a probability is assigned to the unfolding of events, how the utility of a choice is worked out, how all possible alternatives are evaluated.

Otherwise, one has to admit, as does behavioural economics, that when it comes to making decisions, we are often influenced by emotions (Thaler 2015). Instead, the most orthodox economics persists in hypothesizing the existence of rational and omniscient individuals, impervious to emotions and free from any external influence, whose minds work on the basis of optimization algorithms and whose decision will provide the best of all possible worlds. And when imperfect information is introduced—like in game theory—the main concern is more about the possible outcome of strategic interactions among agents than about a thorough examination of the real motivations of individuals.

In any case, in the long run resources can vary. In this case, economic behaviour would be aimed at increasing the quantity and quality of goods and services at one's disposal.[1] Of course, we could achieve this result on our own, counting solely on our ingenuity. But we know that it is much easier when we can nurture and enhance individual skills and form a social organism. *Homo oeconomicus* is therefore better defined as a figure that pursues its own interests within a wider collective endeavour[2] aimed at augmenting the social availability of resources through the division of labour, leaving the problem of their distribution to the political sphere.

A keen reader will realize that we are dealing with one of the most controversial topics of economics—the distribution of the social product to the so-called factors of production. For some, this relationship is determined by the conditions of efficiency of the economic system. For others, it has an exogenous character linked to social conditions of production.[3] After decades, the matter is still unsettled.

---

[1] This distinction between short-term conditions and medium- and long-term conditions of reproducibility was introduced by Alfred Marshall more than a century ago in his "Principles of Economics". He also owns the emphasis on the importance of the "social organisms" that the *Homo oeconomicus* is able to generate. This approach inspires our considerations.

[2] It might seem that this definition reverses the idea proposed by Adam Smith (1776) that the collective interest would be the result of the pursuit of individual interest. But this is incorrect. What has changed in this case is only the unit of investigation. Political economy (a theory of choice) no longer focuses on the individual behaviour of independent agents (as does microeconomics). Instead, it refers to the social organisms that individuals, related and interdependent, manage to form. This latter perspective is the most appropriate for our evolutionary approach.

[3] See Tiberi Vipraio (1999) for a discussion.

During the division of labour, when everyone specializes in what he/she does best, incorporating this knowledge into increasingly detailed tasks, and creating more specific products, available goods and services increase dramatically. We now need a set of rules on how to exchange all the products we make out of those resources and skills. For example, we could reward the effort required to produce them. In this case, we would form more egalitarian societies. Alternatively, we could rank the relative skills of individuals according to their competence and ingenuity on the job and form more meritocratic societies. Or we could pay a premium (rent) to those who claim a right of "ownership" on some resources and form a pyramidal society.

Note that the distribution of the social product is related only roughly to the form of government, which can range from a dictatorship to a democracy, with everything in between, without determining how much of the whole one eventually gets for himself. In other words, even if it is easier, in principle, for a democratic system to provide a fairer distribution of the social product than a dictatorship, this result is not always guaranteed. In fact, an elite could indicate which representatives of the people are to be elected, within a limited set of possibilities, and call this procedure democratic even though, once elected, the representatives of the people will only respond to such elites. Think of the most financially advanced countries of our days. Conversely, an "enlightened" dictatorship could rely upon the popular support of more-protected citizens for some particular end; for example, its geopolitical supremacy in some area of the globe. Think of last century's emerging countries in Southeast Asia. In any case, what changes as a result of the different forms of government is only the form of the social pyramid, which can be more or less pointed while remaining a pyramid.

Returning to our economic issues, we know that the division of labour and technological progress have increased the availability of goods and services and led us to create complicated products that one individual would never be able to realise alone. We live in a world of continuous exchanges. Let's see when this began, and what prompted it.

11.1 When It All Began

We have seen that between 200,000 and 100,000 years ago, the Sapiens had very slowly developed their lithic technology, while symbolism was scarce. Our Sapiens ancestors resembled us fairly closely in anatomy, a little less so in physiognomy, and very little in cognition. They already had a culture, perhaps with the first glimmers of symbolic thought. But they had not yet resorted to the division of labour. Of course, they produced useful goods. And perhaps they exchanged them for other goods. But it does not seem that they could form very large societies as a basis for the graduation to large-scale specialization. What made a difference at some point? First, we asked ourselves how symbolic behaviour generated different cultures. Now we ask the opposite question. What prompted a change in such stable cultures, which lasted tens of thousands of years with minimal variation in tools? What generated so much innovative fervour in the working and embellishment of all objects? And what, if any, other elements came into play?

We have said that the desire to increase our wealth is a typically human characteristic, and that it began when we started to equip ourselves with tools; those who had an axe were richer than those who had to make do with their two arms. But it was a difference that was easy to bridge through imitation. Of course, at some point imitation alone was not enough, and it was necessary to transmit instructions and procedures via a complex language. This language, together with gestures and imitation, needed to allow information to be transferred with high fidelity (Laland 2017). Otherwise, the information would have been lost, especially between generations. But if we refer to the significant increase in material wealth generated through specialization and trade, to when can we trace this initial behaviour? And where do we find the oldest evidence for economic behaviour in the sense we just gave it? We will argue that these traits appeared in modern humans with the emergence of symbolic thought.

According to many scholars, modern human behaviour emerged in Eurasia about 40,000 years ago when the first rock art and portable art began to appear. Others argue that modern behaviour developed much earlier, motivated by artistic and economic factors, not in Europe but in Africa. In the South African cave of Blombos are marine shells of the *Nassarius kraussianus* species, all perforated with appropriate stone or bone awls. Some are blackened through the elaborate techniques of heating or painted with pigments from pulverized minerals (Henshilwood et al. 2004). It would have taken many hours of work to do the job. The shells were held together by strings, now lost. But by analysing the traces of wear on the shells caused by the strings, researchers reconstructed how many knots had been tied to divide and assemble the valves in different styles (Vanhaeren et al. 2013). These objects, however, do not seem to have any practical use. Yet, they have been found all over the continent, even in sites far inland. Nassarius shells of different species, perforated and painted with similar techniques, have also emerged in Morocco, Algeria and Tunisia, dated to at least 80,000 years ago. In the Middle East, similar finds date back to 100,000 years ago. What were these objects for?

It has been suggested that those perforated shells could have been used not only as bracelets and necklaces but also as a currency.[4] They would have been the first forms of money in history (Tuniz and Tiberi Vipraio 2016, 2018). It is a strange coincidence that a metallic coin engraved with the image of a similar shell circulated in Ghana until recently under the name of Cedi, which, in one of the local languages, means cowrie shell.[5] Other mysterious objects found throughout Africa are sticks and tablets with crossed-line engravings. What could these abstract annotations

---

[4]If these ornaments were considered jewels, the authors suspect that "female favours" could be the object of some exchange, either to seduce a female or perhaps to compensate her family for her loss.

[5]In fact, until 1901, coins engraved with the effigy of a shell circulated in that region, together with real shells.

mean? Could it be a form of accounting? Could it be dues to be paid after a number of exchanges? To answer these questions, let us remember what characteristics an object should have to function, first as an intermediary of exchanges and then as a symbol of wealth, set aside for subsequent exchanges.

11.2 What Is Money?

It is said that the main function of money is to be an intermediary of exchange. To perform this function, the chosen object should have precise characteristics. It needs to be difficult to find, because otherwise anyone could create money, and it would cease to function as an "invariable" measure of the value of the exchanged goods, causing what we call inflation. In some cases, conventional objects could be used, applying some kind of complex, recognizable and difficult-to-replicate work. Modern analogues are effigies imprinted on coins or the graphics on our paper money, which names the exclusive issuing authority.

But to discourage the creation of money, apart from the threat of jailing counterfeiters, it would do if the cost of "fabricating" it exceeded the cost of making the goods for which it is exchanged. Alternatively, the work necessary to create money should be of "rare quality" and therefore difficult to replicate. Or we could identify goods whose availability grows slowly, in line with population growth, and use them to act as money.[6] Meanwhile, this object should be easily transportable so that it can change hands readily. Otherwise, it should be "represented" by another object with these characteristics (for example, a reliable promise to transfer the original object in the future). Finally, it should be a non-perishable asset, to maintain its value over time, and then allow subsequent exchanges.[7]

If we refer to these characteristics of money, the shells of Blombos, found throughout Africa, could be considered both as personal ornaments and as intermediaries of exchange. If this were the case, it would be one of the first processes of abstraction in which symbolic thought was expressed: the human representation of a conventional value transmitted and recognised within a community.

The result of this reasoning is strengthened if we consider money in a more modern sense—not only as an intermediary of exchange (a flow), but also as a set of debts and credits (a stock) to be paid in the future. In this case, the currency would represent a certain amount of wealth and consist of a transferable credit instrument. Which object was used as a currency would be quite irrelevant. All that is needed is a system to record these credit and debit accounts. As long as one can count, the nodes of a string or the regular and repeated incisions on a tablet could perform this

---

[6]For example, until recently, cattle and herds were used in some societies in exchange for other goods and services, and even a wife. The stability of this supply of "money" was ensured in this case by feeding the family with the newborn animals.

[7]For a brilliant discussion of these and other aspects of the concept of money, see Martin (2014).

function. In fact, 80,000-year-old ochre tablets engraved with signs of this type were also found in Blombos (Henshilwood et al. 2009). Sticks of wood, bone or other material, often called notched sticks, with similar characteristics were there, too.[8] To function as a "reserve of value", it is enough that someone is willing to exchange goods and services for these objects. In other words, that someone attributes a symbolic value to them.

After digging and ploughing the land for many millennia, symbolic thought was sowing the seeds of modern industrial capitalism. Today, most of the money we use has no physical form. According to some statistics,[9] the intangible component of the currency amounts to 90% of that used in the United States and 97% of that used in Great Britain. Future archaeologists attempting to read the fossil record would be mystified by this aspect of our lives today. But it is possible that some of our contemporaries don't have much grasp of it either.

11.3 Goods and Services

Returning to the real economy, the total well-being of a population depends on how much the exchange of goods and services (obtained through specialization of separate tasks) generates a surplus of goods and services to be exchanged. If this surplus is not created or is not redistributed satisfactorily, someone might find it convenient to produce all that he needs on his own and forget about the division of labour. After all, he could argue, that freedom is priceless: It is better to be content with what one can achieve than to be conditioned by social rules that provide little benefit.

This is not a widespread attitude and is hardly tolerated. Today we live in a society that largely benefits from the division of labour and a specialization in increasingly detailed tasks. These advantages depend on available technologies. The technologies of the past can be studied in the instruments found at various archaeological sites. Technological development is related to changes in environmental conditions and demographic trends. What is the archaeological evidence in this regard?

Between 100,000 and 70,000 years ago, South African Sapiens had developed new stone processing methods, including pyrotechnology, in which advanced two-sided tools were produced using heating techniques (Brown et al. 2012). At the Pinnacle Point site mentioned earlier, small bladelets (microliths) were found that

---

[8]It is amazing how these sticks resemble the collection of tallies from the English Exchequer. They were used throughout the Middle Ages to document the financial relations of the English Crown with landowners and other creditors. They were mostly destroyed in 1826, on the grounds that they belonged to an accounting system far too primitive in comparison with the money issued by the Bank of England.

[9]Calculations made by the Federal Reserve Bank of St. Louis for the United States and by the Bank of England for Great Britain.

had needed at least six stages of processing: identifying suitable stones, collecting firewood, treating the stones with fire, preparing the nuclei (or cores) of the stones and, finally, producing the blades and finishing them in their different final shapes. Their insertion in an arrow shaft or lance required the addition of feathers and adhesives, followed by further processing steps (bending the wood, and interlocking techniques). The final result was a useful *and* beautiful object.

On the one hand the microliths were probably the tips of new, deadly throwing weapons, which used wooden rods as a propeller (as the Australian Aborigines did until recently). Considering the advantage these weapons would have conferred, these results square with the start of the arms race that we mentioned when private (group) property was born along the coasts of South Africa. On the other hand, when we speak of "stone age", we refer only to what is left. Many objects and materials connected to the finds of stone have been lost because they were perishable. We can only imagine their beauty, functionality, and the craftsmanship that went into them. But we do know that the whole process involved work that could last from weeks to months, and the ability to carry out complex operations with precise aims and creative techniques, in this case handed down for more than 10,000 years.

What relationship can we envisage between rapid changes in environmental conditions, demographic trends and technological innovation derived, in turn, from cultural variations? When the former change dramatically, as happened between 100,000 and 50,000 years ago, the effects can be the same as when a war occurs. At first, the population should decrease. Later on, hardship stimulates ingenuity, generating innovative behaviours. This favours a greater accumulation of resources, which then turns into a subsequent demographic increase, if the environmental conditions permit.

It seems that, with regard to the spread of cultural innovations in general, the size and number of the different groups, as well as their degree of interaction, are decisive (Powell et al. 2009). The size of the population, it is claimed, would have reached a critical point 100,000 years ago when the natural losses of knowledge amassed previously would have been more than compensated for. This would have improved survival conditions and triggered a feedback mechanism on population growth (d'Errico and Stringer 2011). At a certain point, when the population became large enough, and a certain division of labour was generated, the first exchange of goods and services should have begun. This point of view is confirmed by some recent archaeological surveys.

The correlation between climate, technological progress and demographic trends was studied at the Pinnacle Point site in South Africa, where one can read in detail the environmental register of that period, which is characterized by the alternation of various glacial stages. In Africa, these translate into periods of drought. In one of these periods,[10] new characteristics in stone technology, including pyrotechnology, appeared, reflecting intense changes in the economic and sociocultural relations of

---

[10]This is the Marine Isotopic Stage 4, dating about 70,000–75,000 years ago, which corresponds to the eruption of the Toba volcano.

the population in sync with the droughts. Evidence suggests that with high climatic and environmental difficulties, there was an increase in the size of social groups, a decrease in mobility in the territory, and an increase in technological development. These are all conditions that favour specialization in distinct tasks within broader social groups, and the accumulation of resources near settlements (Wilkins et al. 2017). An increase in the surplus generated collectively is now associated with an increase in the population, despite the changed environmental conditions, which would affect population in the opposite direction.

To create favourable conditions for trade, however, the various human groups must generate a certain overabundance of available goods, which is not all absorbed by the increase in population. Later, thanks to an increase in specialisation, these surpluses will increase further, spurring a further rise in population. This generates a virtuous cycle that feeds itself. Can we consider this mechanism as triggered merely by a change in environmental conditions? Or does it need some other element, such as a complex language capable of generating new and stronger interpersonal skills? It could be argued that, alongside these two phenomena, another new behaviour has been the catalyst for accelerated progress: that based on the onset of trade itself.

To bring this cultural revolution to fruition, however, it was necessary to overcome a huge obstacle—the inefficiency of the barter system, which would have required repeated exchanges, perhaps indefinitely, before linking buyers and sellers. An intermediary was needed. And if our suspicions are founded, we were able to invent it in really remote times. However, even in this case, as in the arts, we had to wait for the emergence of our ability to generate symbols.

Homo oeconomicus, according to our definition, is now able to emerge into the light after spending millions of years producing only "useful" objects for himself and his close entourage. In order to be truly "economic", and rid himself of scarcity, he must now envision "worlds". He has to give things imaginary values (through the development of money). He has to invent mythological characters (say companies with legal personality). He must imagine which social structures are best suited to organizing, defending and distributing collective wealth (and thus create institutions like states and markets). He has to promote value systems to inspire social ethics (the respect for the law, or the existence of human rights). He might even support those who preach the existence of gods to coerce people into observing the rules. We will now try to examine how this habit of generating an abstract reality began to increase the availability of goods and services, and the total wealth of a society.

The first division of labour is likely to emerge at the gender level: For example, women might consider performing certain specific tasks in exchange for other tasks performed by men, and vice versa. But this is not an economic specialization for commercial purposes. It is about exchanging goods and services free of charge on the basis of trust, emotional ties and any hierarchical relationships. This type of specialization, based on gender, requires a certain degree of continuity of relationships. This is true within every type of family group, and even within small communities. To make it work, one of the following conditions must be met. At best, a condition of reciprocity must be in place (I'll give you this; you'll give me that). Alternatively, when the exchange is asymmetric, and benefits somebody much more than others, a certain degree of power, in whatever form, must be exercised. When these elements fail (for example, when the exchanges are episodic or between strangers) but the benefits of the division of labour remain, it is then that commercial exchanges probably begin. In this case, trade is facilitated by using some particular object to act as money—first as an intermediary and then as an account of all dues established during a series of transactions.

In trading occasionally with outsiders, confidence is low. Now we have to make sure that at least two other things happen: first, that everyone recognizes the same conventional method of keeping receivable and payable accounts; and second, that the object used for the balance of the differences is a transferable credit with a stable value over time, both to accumulate the stock of wealth in itself, and to fuel the flow of subsequent exchanges.

If we reduce the root of our more complex contemporary economic behaviour to these concepts, then we can say that it was born in deep time with the emergence of the division of labour and its application outside the first family nuclei, by means of some form of "money". But to be able to invent money, we needed to draw on our imagination, assigning a symbolic value to some object, and making it recognisable as such to outsiders.

Today, the value of each trade intermediary and of every debit/credit relationship exists only partially in the object we have decided to adopt as currency: a little more in rare and shimmering metals, such as gold and silver, which we could use for their beauty and permanency; almost nil in the charming little paper sheets that we use when we pay cash; and totally nil in electronic transactions. Almost all of us believe that the money we have in our pockets or the credit recorded on our current account has value in itself. But this is a mistake. In fact, we would not know what to do with it if nobody accepted it as a means of payment. The value of a currency, like that of beauty, lies in the eye of the beholder.

11.4 The Private Accumulation of Wealth

So far, we have only expressed hypotheses on how our material wealth has increased due to exchanges paid for with money. But it is unclear how we have created societies where wealth is so concentrated in a very small group of people while others struggle to survive. This is a typical situation of our era; indeed, inequality is said to be increasing in recent years. Without getting into the details, it is worth noticing that the concept of wealth is quite blurred, as it comprises many items and is difficult to evaluate. In some societies, many services are provided free of charge (for example, by the family or village), while in others they are supplied for a fee (according to market rules) and might be quite expensive. In national statistics the first society may look poorer than the second, despite being able to enjoy the same services.

A rough approximation of a nation's affluence is often said to be the Gross Domestic Product (GDP). That is the total market value of all goods and services produced within a year in a certain country. Though being a very popular economic standard, GDP is a flow that has serious limitation when applied (as a proxy) to measure the wealth of a nation (a stock of resources that, together with real income, is associated with prosperity and the general well-being of a population). Not only does GDP exclude many valuable goods and services that are essential for economic and social reproduction (like family work, for example). It also disregards the "opportunity cost" of leisure time, which is a precious item, especially in our hectic contemporary life. And it double counts all damages and repairs. For example, can we really consider a polluting country that poisons the environment, and then cleans it up, richer than a country that does not pollute it in the first place? And how can we forget about the national distribution, among people, of the same GDP and be indifferent to whether it is fair or not?

Better measures of a nation's prosperity are indeed available. One is the Genuine Progress Indicator (GPI) which adds to GDP household production and leisure and subtracts "defensive expenditures" while adjusting for inequality. And in recent decades, the United Nations have regularly provided indicators like the Human Development Index (HDI) and the Inequality Adjusted HDI (IAHDI), in which health, education, gender equity and other social features are combined. Nonetheless, proper wealth measures remain elusive.

Today (abstracting from people's well-being), the matter is complicated when we try to assess a realistic value to the real and financial assets that are not exchanged in the market and rest—idle—in one's patrimony. Finally, wealth is almost impossible to evaluate when if consists of bets (derivatives) with a "notional" value that is so huge it implies the existence of buyers from Mars.

Neither is it clear who owns the planet's riches. Is it the state that controls a territory well-endowed with precious resources? Its elites? Perhaps the companies that claim a right to the resources? Or maybe the "citizens of the world", who want to breathe pure air and drink clean water while limiting the exploitation of non-renewable resources. We will now provide a possible explanation of how we have generated so much social imbalance in the distribution of wealth by concentrating on the accumulation of resources (material or immaterial) for the benefit of some but not others.

According to the definition focusing on private wealth, we can reconstruct a possible sequence of accumulation of wealth. Initially, there are at least four sources of advantage, based on the exercise of power: (1) authority (mental power); (2) benevolence (relational power); (3) authoritarianism (coercive power); and (4) money (financial power). We will argue that they all arose in modern humans with the emergence of symbolic thought.

To these, we add a fifth source of wealth based on a better allocation of resources and call it "commercial wealth". We could also imagine a sixth source of wealth based on the distribution of the benefits emerging from technological advances. We will cover it in the conclusions in the context of artificial intelligence and social networks. We now consider the first five sources of wealth separately.

Initially, in forming larger groups, the first social differences derive from a greater abstraction capacity of some people, who would assume the role of leader. Wealth

11.4 The Private Accumulation of Wealth

would manifest itself in the form of "mental wealth". Some people would begin to tell stories, generating the first abstract realities, the first religions and the first rules of behaviour (the first institutions). These people—those endowed with a high level of intentionality – would form social elites and derive their charisma from having conceived of the best narratives to be transmitted to their followers—the stories that the latter would be more likely to believe, depending on their character and needs. And it would be particularly exciting if these stories were told, acted out or sung in a large natural cathedral in which, in the glow of fire, hundreds of animals seem to gallop on a screen, as suggested by the setting of the cave of Chauvet.

Once, among these narrators there might have been the elders of the group, or the generation of thirty-year-olds (men and women of the time) who had greater experience in the construction of tools, the recognition of edible plants, the use of fire or in providing care and comfort through rituals and magic, which could have important placebo effects. The first inequalities would soon arise: on the one side stand those who tell the stories, dictate the rules and transmit the information; on the other side are those who listen to the stories, observe the rules and receive the information. Today the former social class is represented by the intellectual elites, who determine the cultural and technological trajectories of the future. The latter comprises laypeople.

Secondly, wealth might have accumulated for certain people and their families in the form of durable goods, or services, which "represent" subjection, respect or gratitude, and confer a higher social status on the recipient. They consist of gifts, such as ornaments, instruments or other conventional goods dear to a certain culture. Such objects could be rare or, much more frequently, produced with the application of a certain amount of specialized work, as in the case of the thousands of finely worked mammoth pearls found in the tomb of Sunghir.

This second type of wealth could be accumulated in at least two ways. The first is because someone proves to have a greater capacity or works hard in some field, and, in this case, is the source of his or her own wealth. Alternatively, because this person is particularly generous to his fellow men and forms bonds and elicits feelings of gratitude from other members of his group. We could define this type of wealth as "relational" in the sense that it is generated in the asymmetry of relationships. We will see that it can also accumulate in the beneficiaries when they do not feel compelled to reciprocate the favours and take advantage of the generosity of others. In this case, it would be based on the exploitation of donors by their beneficiaries due to some individual or social weakness of the former, or ingratitude of the latter.

A third form of accumulation of wealth has undoubtedly been nourished by the direct exercise of force (authoritarianism). It is the one we study most in the history books: the most skilled, violent or valiant and the most able to fascinate, subjugate the weakest and dominate them. These people impose on others behaviour that increases their own material well-being (securing pre-emptive rights in the most disparate situations, from the harvest of the wheat to the first night of brides). They compete with other characters of their own kind to ensure ever-greater pre-emptive rights. Sometimes, these skills are transmitted to their descendants but sometimes they cause their premature death. Some of these lords and ladies leave monumental works to ensure that their memory lives on.

With the invention of money—a symbol capable of preserving value over time—the accumulated wealth tends to generate more wealth, both for those who borrow it (if they use it in productive activities) and for those who lend it (if they ask a price for the favour). Social inequality tends to increase at the expense of those unable to access this virtuous cycle of reproduction of wealth; that is, for those who are "less reliable".[11]

Wealth is also increased by the division of labour and by technological progress, since a surplus is soon generated that exceeds the needs for survival. When different social groups come into contact, if they produce surpluses of different natures, each group will tend to cede the more abundant goods and services in exchange for those that are scarcer, generating more wealth: this is what we call commercial wealth. A conventional object is needed to make it last over time (perhaps starting with some beautiful pieces of finely worked shells, like the shells of Blombos). But we cannot say anything a priori on how this additional wealth is shared among parties; everything depends on the relative prices of the traded goods and on how the benefits of the exchanges are shared. Even so, we can still admire some of its uses in the modern era, walking through San Marco Square in Venice.

The accumulation of wealth typical of the capitalism of our day is merely a special case of what we have called "relational wealth", which depends on an asymmetry in the debt and credit relations between people. Capitalists simply pay workers a wage that is lower than their contribution to the creation of value. And that is because they claim a compensation for the provision of machines to work with and the organization of production.

However, if we believe, in line with the Marxist tradition, that the capital accumulated by someone (as a stock)[12] is nothing other than the fruit of the work done in the past by someone else (the workers who built the machines that the capitalists now claim to own), the accumulation of wealth takes place due to the failure to recognize this original obligation. The alleged existence of a "title of property" is nothing but a stratagem based on symbolic thought that allows the expropriation of some rights of access to the resources earlier provided by the workers. It is thus based on the cultural and political system of that society. And managers (not shareholders) are highly compensated for their "organization of production". Indeed, they are included in the national statistics as "labour".

---

[11] This effect could be avoided if we chose a perishable item as temporary expressions of wealth. In this case, there would be no incentive to hold it for too long, and everyone would be eager to transform it into "effective demand" and use it as a means of exchange for real goods and services. A disaster for the world of finance!

[12] Note that we are not talking about the flow of investments capable of generating new wealth in the future. If we were, our reasoning would become more complex and call into question other considerations related to the distribution of income to all factors of production (and therefore to the organization of work, and the risk taken).

11.4 The Private Accumulation of Wealth

We must admit, though, that this is a fairly obsolete view of the mechanisms of appropriation of value today. To get rich through asymmetric relationships, it is no longer strictly necessary to exploit the value that the worker has contributed to, although it is always possible, and continues to happen on a large scale. In a society that gives more and more value to information and uses digital networks, it will be enough to technologically control the points where the information is collected and distributed, and then claim the right to appropriate its benefits from those who want to use it. The generators of information (everyone who uses digital technology in their homes and offices) are often unaware of creating commercial value and will be happy to give away a precious asset simply in exchange for a connection service. And the information collector and distributor, presiding over a technological platform, can easily transform a vast collection of information into something that can make him very rich. For example, we are even happy to pay a small fee for the analysis of our DNA—giving away the right to use the results—without realizing the immense commercial value of these data.

All the digital platforms of our day generate for their owners wealth of this kind: a value based largely on an information asymmetry. This introduces a new entity into our story: machines equipped with artificial intelligence. Our fascination with them prevents us from seeing an incontrovertible fact—that most of those in the framework are those who "steal" the work of other human beings (together with us, the digital users). Much more so than those who, for the time being, use only their arms and minds, and perhaps appear at the horizon coming from distant nations.

Every time machines can do something better than humans, unemployment will be created[13] and the benefits generated by the machines will all go to their "owners", unless we are prepared to reduce the working time of current employees and preserve their purchasing power. Wealth will therefore continue to increase and be concentrated in the hands of a few, due to the fact that they now control both machines in the strict sense (direct productive capital and related technologies), wealth distribution (through credit), and the information that we give them for free, through intelligent machines.

A new society of hunters and gatherers (of information) is emerging and available resources appear scarce—for many—only because they are unevenly distributed. Very soon, if not right now, we will have to deal with the changes in our social body and worry about how to handle these changes. Before our social fabric gets torn, it will be up to the elites to decide whether to imagine new stories to tell or simply pass on the old ones.

---

[13] The idea of a basic income for all facilitates this transfer of work to the machines by reducing the social cost of structural unemployment.

11.5 The Destruction of Collective Wealth

Our economic voracity, aimed at concentrating private wealth in the hands of a diminishing circle of subjects, has many advantages for the subjects in question and for those who can enjoy the benefits of a wider society, more informed and connected, thanks to intelligent, omnipresent and seductive machines. However, there is a high price to pay.

According to calculations of the Global Footprint Network,[14] a research centre studying our ecological footprint, the day we consume all the resources that the planet manages to regenerate in a year (the so-called Earth overshoot day) comes earlier every year. In 2019 it fell on 29 July. After this day each year, we go into debt at the expense of our descendants. Plants, animals, clean air, and fertile soil—we are rapidly consuming our endowment from an evolutionary history lasting more than 3 billion years. The problem began long ago when we started to modify entire ecosystems and caused the extinction of entire species, wreaking havoc. And it continued until a few decades ago, when we reached a balance between renewable and consumed resources.

Since then the deficit has been growing. It is estimated that we would need a planet and a half to produce the resources needed to sustain humanity's ecological footprint. And, according to a conservative projection, three planets will be needed before the middle of this century. Time is running out. We must find a solution. The protection of the environment becomes, at least, our moral responsibility. And this even without revolutionizing our vision of the world and continuing to remain anthropocentric (Bergoglio 2015).

Meanwhile, we might reflect on the ambivalence of human nature as reported in this book. This means looking at ourselves not so much as what we believe we are when we place our species at the centre of the universe. Nor can we trust any ideal construction which inspires our actions and which we then assume to be true only because it is appropriate or desirable. The human nature of reference should be, whether we like it or not, what emerges from the kaleidoscope of behaviours observed in the course of our evolutionary history.

The excesses that symbolic thought can bring along when it pretends to have no limits lead us to reflect on the interaction we wish to have with the different aspects of our nature. In theory, we could decide if we want to continue to give precedence to the objectives of a *Homo oeconomicus* who tends to make the most of available resources and to infinitely increase his or her well-being. Alternatively, we might choose to introduce other criteria to socially organize and survive on this planet.

The time has come to identify some possible scenarios and some critical points that lie in the years to come when a new character—artificial intelligence—will enter fully into the plot of our story.

---

[14]http://www.footprintnetwork.org/en/index.php/GFN/page/world_footprint (last access in July 2019).

Chapter 12
Humans of the Future

In dealing with the possible future of humankind we introduce a new character: artificial intelligence. We touch upon the digital revolution and how it latches onto our biological and cultural evolutionary process. In examining the many facets of social networks in association with the self-domestication hypothesis we offer a discussion on the relation between power and pleasure, truth and post-truth, intelligent weapons, and preventive war. Finally, we debate alternative views on our evolution and point to the risks of a possible hybridisation with intelligent machines.

Applying the biology of populations—in the life cycle between prey and predators—to the evolution of wealth and inequality of "civilized" humanity, some have noted the repetition and the alternation of two cycles, one long and one short. The first, which lasts about two or three centuries, begins with more egalitarian societies and evolves into an increasingly inequitable distribution of wealth, the reduction of available resources and mounting social turbulence leading to its final collapse. The second, lasting about half a century, sees the alternation of two generations, one more peaceful and one more confrontational (Turchin 2003). According to these analyses, the year 2020 could coincide with the most turbulent and confrontational phase of both cycles.

To understand and, possibly, intervene on the negative effects of these phenomena, we would need to connect the most important variables of the human system—demographics, inequality, economic growth and migration—with those of the earth system—biodiversity, eco-fragmentation, resource consumption, pollution and waste management—to identify their complex reciprocal interactions (Motesharrei et al. 2016). Otherwise, we risk major catastrophes generated by the progressive destruction of resources and a reduction in technological and social complexity, as occurred many times in the past (Tainter 2003).

That is, if no major changes had occurred in our recent past. However, what if a "revolution" had taken place before our eyes, and we did not take much notice, distracted by the apparent simplicity of its appearance? What if we had stepped into another platform in which we can enhance our previous experiences and *modus operandi*? Looking at our most recent behaviours, so deeply and widely intertwined with digital networks, one can think that we have created a sort of Ultraworld—an augmented reality—that we not only imagine, like in the deep past, but in which we can really live. How did this happen? And how is this going to change our future?

12.1 Digital Networks

Let's start by looking at the pervasiveness of digital technology characterizing the present. By combining our natural intelligence with the artificial intelligence of machines, some claim that we are designing a new, post-modern Sapiens able to circumvent natural selection: a freer, healthier and more long-lived human who lives in an ever wider and interconnected society. In reality, this optimistic view of our future—much in vogue until a few years ago—now begins to lose much of its appeal. But before throwing the baby out with the bath water, let's take a look at what might have prompted the digital revolution in the first place. Was it pulled by technology? Or pushed by culture? Probably both.

The contribution of information and communication technologies (ICT) to our wellbeing has been discussed extensively, particularly at the end of the last century. In those days, ICT were said to induce a dramatic reduction in the costs of production and trigger exponential economic growth. In the past, only the division of labour could produce a similar outcome. Technically speaking, the reasons were found in the existence of three (observed) rules of thumb that functioned as multipliers of value. First, the computing power of microprocessors was seen to double in a period of approximately 18 months, thus halving the bit cost/price every year and a half (Moore's Law). The only limit to this law was the miniaturisation of components. Second, as far as supporting technologies were concerned (cables, fibres, waves), the bandwidths available for ICT tended to double every 6 months (Gilder's Law). Finally, the value of a network was roughly proportional to the square of the number of users (Metcalf's Law). In the competition among standards, only those who were able to reach the higher initial value of users were destined to win. Technology was there. The plane was ready to take off. But the destination was unclear. The digital revolution had to be pushed by culture.

The contribution of an alleged cultural revolution to our current way of life, when permeated by digital networks, is more difficult to assess. At its onset, in the roaring Sixties, there was probably a wish to get rid of all the old institutions that were blocking the right to free expression and movement. The effort could be seen as a sort of insurrection led by visionaries who wanted to overturn the previous order of nations and states, break social barriers, upset and disrupt information enclaves, crumble old rents in transport and communications, music production, media and entertainment, bypass old parties and give people a voice (Turner 2008). Economic motivations, though present, were not paramount, at this first stage. Nobody knew in advance where we were going. For digital innovators, the world was a "trial and error" playing field. In the last decade of the last century, a dot-com bubble developed, and then burst. One had to wait a decade for big money to pile up, but only for the happy few. And how inspiring were the newborn billionaires for some people!

Anyway, what is a digital revolution in the first place? To put it bluntly, it is a very simple procedure. It is the capacity to translate all information in a list of zeros and ones, and thus name, store, transport and share at high speed, and virtually no cost, everything we wish to digitalise. A language was invented to talk to the machines, as they could be made sensitive to the switch (on and off) of a signal. The resemblance with the great leap forward generated by symbolic thought is striking. But with a difference. Instead of representing the world, and making it socially relevant if we all believe in it, we can now represent the world as we know it (actually, as we feel it), and live in it. Digital technologies have allowed us to jump to the other side of the looking glass. We are not passive spectators any longer. After a couple of decades of incubation, with the invention of the smartphone—a telephone, a computer, a window to the world, a friend – we have become active players of a big Game (Baricco 2018). A mental revolution is set in place. We can shortly summarize its main features.

Previously, information was scattered and privately managed and stored. By digitalising all information—"all" is the key word here—we can bypass hierarchies and connect things and people at unprecedented scales. No more ideologies and trickled down ideas. We can forget about old leaders and make ourselves relevant. There are no more elites to respond to: inventing stories and programming behaviours upside-down is outdated. We can overflow the net with millions, billions, of users' preferences and create a bottom-up inverted hierarchical order: a weird situation in which "quantities" rule and finally people get what they really want. There are no physical limits to our experiences. We can now hybridise with our smartphone, and live a full and rewarding life. All we need is the capacity to live in both worlds—the natural and the artificial—and then get the most out of them.

Social relations can be light and easy. Instead of "changing the world", we can generate many worlds, all beautiful and diverse. And with the help on an App, we can create a hangar where we can store them, and experience them. There are no more parents to respond to. Together with our parents, we can be children forever. We lose depth in reaching out to the world, but we gain scope. We can span in all directions. Indeed, we can play a mega video game that develops at different levels. There we can score and compete. There we can win.

In a nutshell, by combining technology's pull with cultural push, another self-enforcing mechanism is activated. One that is making our deepest dreams come true. We could become carefree and fully develop our perpetual childhood. Instead of working long hours and submitting to boring tasks and heavy discipline, we could cheerfully surf the net and always live on a Sunday. No wonder we sometimes think we are turning into God, who finally rests on the seventh day of creation. We can dematerialise our physical reality and make our surroundings cleaner, safer and simpler. We can curl in a nest made of two realities: the "natural" and the "digital". Indeed, we can live the best of both worlds, because in that nest there are no shadows, no imperfections and no complications. We can be part of a liquid society and move around by following the tide, by being the tide.

To the fifth dimension generated by symbolic thought (a world of art, myths, religions, ideologies and institutions), we are now able to add a sixth dimension to our existence: the virtual reality of our digital life, and the possibility to move from the physical to the digital sphere, and back, in less than a second. Previous technology breakthroughs did change our lives. But we had to wait before they changed our minds on a grand scale. Now, we wait no more. Due to large numbers and unprecedented speed, cultural change occurs in a snapshot.

Certainly, some will find this vision overly optimistic, and perhaps others will find it too bleak. Let's just touch upon a few possible shortcomings of this idea from the users' viewpoint. More will be settled further on, when a second wave of "developers"—sometimes the same persons but older and much less naïve—will firmly establish themselves on the net.

First, it cannot be denied that digital technologies allow everyone to access more information and make more informed decisions than in the past. But this is true only if the spread of information is associated with an in-depth knowledge of how this information is generated and disseminated, and how reliable it is. Moreover, digital technologies not only provide information "to users", but also "on users". In fact, they monitor the way and the forms in which individual liberties manifest themselves, all the way down to the most intimate behaviours. It has been pointed out that, when we look at our smartphone, it is the smartphone that is looking at us. But we don't seem to be concerned and gladly oblige.

Second, political and economic institutions can exert greater control over the population; statistics can evolve into Big Data science, creating individual profiles and predicting people's behaviour (Han 2014). With digital technologies, especially social networks, we can create emotions and then secretly exploit them commercially on a large scale. By increasing the complexity of human relationships, digital information makes society more vulnerable. The more complex a system is, the more it is exposed to shocks. In short, greater freedom of information and communication can go hand in hand with greater control over the exercise of this freedom. But whether more freedom or greater control will prevail will not be determined by the technology itself. As always, it will be up to us to decide how to use it and the risks we are willing to take. The two tendencies are not antithetical. We could attain more freedom to express and inform ourselves about many personal issues (as we do on Facebook and other social networks) and greater degrees of control on issues that affect the community, in particular security or the right to privacy (as happens in the Chinese Social Credit System).

Third, alongside these general aspects, other problems might emerge. We have seen that with the invention of writing and then all other means of information storage and exchange, we have overcome all the temporal and spatial barriers that had retarded the accumulation of knowledge, the diffusion of ideas and interpersonal communication. Of the more than 7 billion humans now inhabiting the Earth, nearly 60% have access to the internet and two-thirds use mobile telephones.[1] In all the

---

[1] https://www.slideshare.net/DataReportal/digital-2019-global-digital-overview-january-2019-v01.

previous phases of our history, we had improved the connections between our neurons and the connections with other people. Now, we are improving the connections with intelligent machines. We would seem destined for endless progress. But is it really so? Indeed, the feedback mechanisms of the new digital technology on our brain do not always favour human sociality (Greenfield 2015).

Finally, even detaching slightly from the rosy picture depicted above, the advantages of digital technologies in enhancing interpersonal communication were generally thought to exceed the disadvantages of declining use of some brain functions, such as memory and analysis. The gains from communicating with the multitudes seemed to outweigh any disadvantages caused by "noise", such as the transmission of photos of a pizza or our opinion on some inane topic. Is it possible that this "net benefit"[2] of communicating on a global scale is now coming to an end and that digital technologies are reducing our "wisdom" as a species? Let's restrict our consideration to the most popular uses of digital technologies.

12.2 Social Networks, Digital Games and Search Engines

It is now established that digital social networks, if central to somebody's relational sphere, can reduce face-to-face communication skills and empathy. As in most of our habits, responsible use could be harmless; it is abuse of the system that damages. By communicating mainly through a text or image to an unknown audience, we can no longer get feedback from facial expressions and body language. For millennia, we have used these signs to hone our communication skills, build trust and distinguish between friends, acquaintances and foes. The lack of face-to-face contact leaves us in the dark about what others think of us; we do not even know who will eventually receive our messages. And the physical distance combined with the perception of anonymity disinhibits us, making our language more abrupt and aggressive. It also leads us to give information that we might otherwise have kept to ourselves.

Furthermore, personal identities can be artificially constructed and do not reflect who we are but who we would like to be. Each behaves as if he/she were at the centre of a stage, playing the lead role. By encouraging the latent narcissism in us all, at least three negative effects might come into play. The incentives to achieve personal, relational and professional goals might weaken. Possible difficulties in socializing could be accentuated in the most introverted, pushing them to over-develop their digital personality to the detriment of the physical one. Finally, serious relationship problems could arise in the world of physical communication, when one's digital

---

[2] A comparison of quantity and quality of information transmitted through digital technologies is very problematic. Even a huge amount of transmitted data can be of "high quality". Think of what happens inside the CERN in Geneva during every working day. We want to emphasize that the increase in information transmitted through digital networks can correspond to both an increase in useful knowledge to enhance our understanding of the world, and to background noise based on a continuous exchange of irrelevant news and tweets.

identity is very different from the real one, generating disappointment and rancour, when this fact is discovered in a physical encounter.

As for digital games, they can provide a feeling of instant excitement through the release of adrenaline and dopamine and improve some intellectual performance, such as response time to visual and auditory cues, and multi-tasking ability. This is good if one wants to become a fighter pilot. But playing long hours favours a tendency towards riskier behaviour, a reduction in attention span and heightened aggression, to the detriment of the analytical sphere of reasoning.

Finally, relying too much on search engines to obtain reliable information, and surfing the net, help the most agile but also less-focused mental processes. We also risk darting from one piece of information to another, confusing and conflating ideas, making us lose sight of the initial objectives of our research. And this is regardless of the reliability of the information provided by the search engines themselves (in some cases, quite low) and the increasingly sophisticated disinformation techniques that can contaminate many sensitive topics, ranging from personal safety to collective security.

The so-called digital natives are at risk of becoming more like machines in that they might grow insensitive to the emotions of others. But they are seemingly more "human", with all their frailties in full view. Digital natives include people seeking attention as individuals but who also adhere to a collective identity and shared mentality in the various tribes of the network. Digital communication amplifies feelings and emotions and gives rewards capable of creating dependency. It also causes desensitization to the feelings of others because empathy generated by physical communication is lost.

In short, the new digital technologies allow us to push our human drives to the extreme, freeing us from the constraints of our bodies and relieving us of a sense of responsibility for the irreversibility of our actions. By learning to present ourselves without inhibition or in a captivating way, we can extend the number of our "friends" and be "someone" even if we do not have the talent and status in real life. Even our desire to belong to some tribe can be easily satisfied, for it doesn't involve any effort and membership is easily cancelled with a click.

The infantile wish to be seen as special can turn into extreme egotism. In general, the desire for instant gratification is cultivated in cyberspace, where actions have no long-term consequences. It is not a world that encourages us to reflect on the past or worry about the future. Yet, our tracks are hard to erase and can be used against us. These considerations apply above all to those who live mainly in the digital world. Fortunately, there are many digital natives who continue to live on this side of the mirror, and can create satisfying face-to-face relationships, perhaps using social networks in moderation.

During our long evolutionary history, we have developed different minds that can connect with each other in many languages, generating collective identities. These minds emerged from the neural networks activated by our individual experiences. They were, and continue to be, different because they also depend on the relationships we establish with people and the environment. Communication took place among bodies that were also different, and never perfect, something we willingly accepted. Our identity was the fruit of our history, a true story proceeding through trial and error, for which we paid a price. Our learning also took place through our bodies.

The timeless and spaceless connectivity we are experiencing now could radically change this perspective, making us forget who we really are and how to relate to the physical world. We could then become entangled in a dense network of networks populated by unreal characters. These would be the result of our imaginations, but also of the new technology of "augmented reality", which can enrich (or mask) the world generated by our sensory perceptions with information created by digital algorithms. Sometimes this happens just for the fun of it (like in the case of the mobile game Pokémon Go). But in many cases algorithms provide information that is very useful: for example, while touring a city or repairing someone's heart. This augmented reality is expected to have a huge impact in important sectors, such as tourism, design, entertainment, medicine and military activities, to name a few.

12.3 Market-Driven Digital Gurus

How come a revolution, perhaps an insurrection, born with the intentions of discarding old oligarchies and freeing people from spatial and institutional bondage, has become a terrain that now reminds us of a minefield? History teaches us that, sooner or later, every revolution settles down and generates new elites. Sometimes an autocracy replaces heyday fighters and extinguishes people's dreams. Or the utopian idea of a level playing field (the basis of free markets) is replaced by the dominance of large transnational entities. What happened to the digital revolution? The usual. Once beat, or silenced, the initial enemies lost importance (at least temporarily) and all efforts were geared towards a higher efficiency of the initial ideas, driven by large numbers and strong multiplier effects.

The fact that old oligarchies did not quite understand, in this case, the deep roots of the process, and agreed to its development, certainly helped the digital revolution to affirm itself. In the meanwhile, under certain conditions, new rents could be extracted from the demise of the old ones. Behind the curtains, there was a brilliant intuition of our deep nature. People wanted to play all along and be forever young.

If we look at our recent digital devices, for example our smartphone and all the icons that it contains, a few striking features emerge. The solution of our problems can be enjoyable and even funny. The look of our "helpers" is very cute and reminds us of kindergarten. We are substituting the old symbols (which require a common vision of the world to be shared and endorsed) with icons (which only involve association of meaning with a clear and unmistakable object: a telephone receiver, an envelope, a book). And last but not least, we can attain outstanding results with a simple touch, just like children, and forget about all the effort that is behind it. The ethics of accumulating knowledge and awareness through hard work falls to its knees. Indeed, we are hybridising with that device with joy. We can now free our experience from fatigue and live on the surface of the world, surfing any wave; actually, making waves.

Of course, there is no such a thing as a free lunch. We are also becoming aware that some people are getting very rich from this world. And that when something (a product, a service) is totally free, "we" are the product and the service. So even if instruments are nice and easy, the Game is difficult. Here is a list of a few emerging features that start to worry us.

First of all, only very few big players have survived. In the last decade, they have been swallowing all small players, and making alliances. Indeed, they are buying all innovative ideas and are the sole investors in artificial intelligence. A lot of profits (rents) are made from collecting and organising the data we drop on the net. Big players pay little taxes and thus lean on those who must pay them. A market for ideas, for news, and even for the "truth" is up for grabs. Former developers of goods and services are disappearing. Increasingly, money is made out of distribution rather than production. The social pyramid of income (and wealth) is rising to unprecedented levels, but only for a few players.

Again, perhaps, this picture is too dreary and forgets about the benefits. We thus leave it to the reader's judgement. In the meanwhile, returning to our fifth dimension, we will see that a number of mines have been lodged there by the digital revolution, and wait to be stepped upon.

12.4 Socio-Economic Networks

In this book, long before we entered Sapiens' latest achievements, we wanted to retrace his history within the larger human family to which we belong. An important aspect of this narrative concerns the relationship between *Homo sapiens* as an individual, and *Homo sapiens* as a social organism dominated by the rules of coexistence and exchange. When the social sciences deal with this topic, the main concern is to reconcile the analysis of individual behaviour with that of the community. Economics is built mainly on this basis, starting from the individual,[3] and then assembling individuals into a social body. If everyone pursued his own interest, said Adam Smith, "under certain conditions", the best of all possible worlds would be generated. But then we forget what these conditions are, namely that we should compete on equal terms. In this case, we would live in a transparent world in which no individual had more power, or more knowledge than another—a world very

---

[3] We refer to neoclassical economics, always concerned with its "micro-foundations" when it refers to communities. Political Economy, instead, sprang from moral philosophy, asking itself how the wealth of nations was formed. The answer was: through the accumulation of money (mercantilism). This position, which remained latent in the contemporary world, is now back in fashion. It is usually considered wrong from a theoretical point of view. But it is justified if we refer to the considerations made here regarding the symbolic role of money. See Tiberi Vipraio (1999) and related quotations for a thorough discussion.

different from ours. To analyse collective performance, conventional economics simply sums up the behaviour of individuals as if the rules of coexistence were given and immutable, as if governed by the principle of rationality. The idea that other principles might come into play, for example ethical considerations, is normally disregarded, even though strategic considerations might be introduced to study the result of interrelated decisions (for example in game theory).

The need to consider the economic behaviour of a society as a whole—the way it is, rather than the way it should be—is now gaining ground in the social sciences, not only in its own right but also thanks to the most recent advances in biology and neuroscience. Certain individuals or intelligent machines could dominate by controlling the dissemination of information. And so could large industrial companies with global networks (Vitali et al. 2011). Financial institutions have a significant influence on collective choices, no matter how developed a democracy. Some could have better or faster access to information. So, while the "children" are having fun exercising their freedom of expression in social networks, the "adults" will take responsibility for the things to be done.

This perspective coincides with the hypothesis of self-domestication derived from the preservation of youthful characteristics in adulthood, with the psychological submission of the tamed to some authority figure. It is also consistent with what we observe in contemporary societies, where the rise of authoritarian populism suggests that we can no longer assume that liberal democracy is the wave of the future. Indeed, it seems to stand at a critical point. The forces of technology, political economy and social identities seem inclined to tear apart the political systems of the past and push what is left of them in two opposite directions. On the one hand there is an extraordinary growth of illiberal democracies run by demagogues; on the other hand, there is a consolidation of undemocratic technocracies, governed by unaccountable elites. Fixing the economy and renewing civil faith are said to provide a good starting point to invert this trend, within a domesticated inclusive nationalism (Mounk 2018). But it will hardly suffice, in our opinion.

In both cases, there are big disparities in the appropriation of collectively produced wealth. Under our definition of *Homo oeconomicus*, and contrary to mainstream economics, the well-being of individuals depends on choices that compete in the political sphere. Many social tensions can emerge from this sphere. And like a huge house of cards made of small pyramids, our social structure could become very unstable when the ideas that keep it together start to fade out. In what follows we shall look at the formation, and stability, of a number of pyramids.

12.5 Social Networks of Knowledge

If we look at how the Higgs boson particle was discovered or how we got to the CRISPR method of genetic editing, it is immediately obvious that scientific knowledge and its applications result from a collective effort. But we believe that this is a characteristic of the present and that, in the past, individual "great minds" were behind the main leaps forward.

Without detracting from the genius of Leonardo da Vinci, Galileo Galilei or Albert Einstein, it should be noted that it is always a social organism that provides the broth of cultivation of each new discovery, just as it is always a social organism that allows new knowledge to percolate and spread. Not only do we use the knowledge of those who preceded us, but we collaborate continuously, sometimes without apparent results, in exchanging hunches, ideas, methods and observations.

The same applies to the knowledge of our daily lives, which is made of innovations that increase our individual and collective well-being, and to what economists call the expenses of "research and development". Combining these two concepts, however, it is quite complicated to understand the importance of the social component of knowledge. Investments driven by intellectual curiosity and uncertain applications (basic research) are mixed with those aimed at obtaining a result (applied research) and others related to their possible applications (the development of products or processes).

In this cauldron, it becomes difficult to assess the social contribution and the collective impact of each expenditure. In short, the social system of knowledge, if applied to the economic sphere, generates, again, hierarchies, which are hidden by bundling up, in a single heterogeneous category, resources that have so little in common apart from having to be financed. A new "knowledge-based economy" would require analytical techniques and theoretical approaches which exceed the current economic orthodoxy based on the obsessive search for the conditions of equilibrium of linear systems. This quest has high ambitions of theoretical elegance but little taste for practical relevance.

Another example of how complicated it is to study social networks of knowledge is the interaction between work and technological innovation. A lot of work is needed to find out how available knowledge can be used, what knowledge is needed and how to make it effective, durable and systematic. Once a virtuous circle has begun between available knowledge and a successful selection of relevant knowledge, it is difficult to value knowledge itself. The advantages deriving from an early exploitation of it become permanent and irreversible. We are bound to perform continuous improvement in processes, products, services. We are tricked into a never-ending Olympics. In a nutshell, to incorporate the interaction between work and knowledge, we need to put innovation and knowledge at the centre of the dynamics of the production of wealth. This perspective should allow for the formation of hierarchies, and distance itself from the ideals of perfect competition, in which benefits are spread evenly. They are not. And we must admit that, if we want to be prepared for a possible rise in social instability.

Looking at geographical areas and considering the existence of states and their tampering with "free market forces", the disorder increases further. When the ability to apply and exploit knowledge becomes the key factor for the success of a nation—and therefore the strategic element in a country's competitive position—"know-how" catalyses economic growth, widening the gap between those who have entered this virtuous circle and those who have not. In our narrative, all these problems can be summarized by saying that we are part of a complex hierarchical organism consisting of a "network of networks" or, as it has been described, a "system of systems". To understand it, a so-called "complexity approach" is now beginning to appear in economic and managerial studies.

12.6 Global Networks and Territorial Networks

With a complexity approach, even political science would need to change perspective. Let's consider a possible conflict between the need to defend our local relations with other people and that of digitally connecting on a large scale. Until recently, it seemed that the latter trend (globalization) was prevailing over the former national and sub-national social structures. Indeed, the two reticular structures are now in open conflict. Each of us lives on the square of a chessboard formed by the various territorial entities we invented. But at the same time, we're linked to a complex web of supranational relationships.

With the digital age and fast transport, this picture has undergone radical transformation. For about 20 years, at the turn of the 20th century, global communication networks were superimposed on territorial networks without causing clear conflict. At the end of the last century, the number of states almost trebled as a result of the political disintegration of Eastern Europe and central Asia and others might emerge under the ethnic autonomist drives of many national states. World governance is now split between a chessboard of states and a web of transnational digital connections (social, economic and political).

On the other hand, we are entering the 5G era, a new Wi-Fi broadband standard 100 times faster than the existing ones. The signals will travel much quicker than signals in our cerebral networks. The web will also connect "things" and we'll enjoy smart homes and smart cities. We will live inside a network of phones, computers, cars, appliances and intelligent wearables. It is estimated that by 2025, the number of objects digitally connected, now standing at 7 billion, will explode to 100 billion. The digitization of industrial production and services will further spread, ranging from entertainment to remote surgery.

Global networks also operate with aims and methods often out of state control. It is along these networks that terrorism and organized crime can flourish. Politics and business can make deals in a grey area. Besides, digital technologies are not alone in being transnational. Think of climate change and environmental disasters, displacing people and animals. Their effects are global and require a global response. If states keep reasoning as if they were squares on a chessboard rather than nodes of a wider network, it is difficult to know how to give adequate answers to questions on the nature and extent of many phenomena. In the face of massive migration, it is much easier to erect walls than to manage the large network disruptions that caused the exodus in the first place.

According to the United Nations, between 2008 and 2015, environmental disasters caused the flight of 203 million refugees.[4] Rising sea levels might generate an additional 2 billion refugees by the end of the century (Geisler and Currens 2017). These are fairly rough estimates. We do not know exactly when all this will happen, because environmental changes will not be gradual once a "breaking point" is reached. At these critical points, the phenomena become cumulative and unstoppable, combining their effects with those resulting from increasingly scarce resources. As in the past, we need to identify shared rules to govern change. Is it possible that the solution of such a complex problem is in the hands of the people, and that they know it?

We are facing a serious institutional problem, the solution of which will be very important for our future. Some propose to develop new strategies that integrate the government of states with that of the regional and global connections in which they are inserted (Slaughter 2017). The new geopolitical strategies should overcome the old political and geopolitical distinctions and explicitly refer to the clash between open and closed systems. It is argued that a global strategy, based on three pillars, should be promoted: open societies, open governments and an open international system. This scheme would be based on broad civil participation, high transparency and accountability of existing institutions, and broad public access to digital information.

According to this point of view, a more open geopolitical structure would be much stronger and resistant to external shocks. Unlike a hierarchical structure, complex networks present an increasing self-organizational capacity when external conditions change. But to do this, the systems must be open. Moreover, the states should operate both as squares on the international chessboard and as nodes on global networks. To do this efficiently, we should increase the transparency of the networks and accept the idea that a state cannot always guarantee the absolute security of its citizens: a minor sacrifice—it is claimed—with respect to the loss of freedom.

Unfortunately, this is exactly the opposite of what is happening. The danger of a cyber war between different areas of the developed world is no longer remote. The news is full of reports of states that penetrate the information networks of other states, even those of allied countries. Some help spread false information or amplify the contents of messages for or against certain initiatives. These are but illustrations of a type of war that will preoccupy us for many years to come. Through appropriate algorithms, it is possible to segregate the world's population into thousands of subgroups on the grounds of politics, religion, tastes and opinions. It is then easy to work out what issues these groups are most sensitive to, and therefore open to manipulation. With the help of AI algorithms, the dissemination of well-constructed

---

[4]http://www.unhcr.org/protection/environment/5975e6cf7/climate-changedisaster-displacement-overview-unhcrs-role.html (last access in December 2017).

messages can make a difference in elections. This is said to have happened already in the 2016 U.S. presidential election and in the Leave.UK campaign.

The global communication networks could also shatter into nationally supervised and protected networks in which the states would have the right to control all real and digital communities (Cooper Ramo 2016). There is even talk about controlling citizens' access to the digital world, and the transformation of the current open global networks into closed networks, the so-called gatelands, protected by digital gates and walls.

Is it possible that some of our evolutionary traits—our ambivalence regarding mobility and stability, competition and cooperation, freedom and control—are now leading us to form a new chessboard made of ever-smaller and disconnected squares? It seems that in many parts of the world this is what people want. Yet, in our long history, we have always managed to progress, thanks to our mobility and adaptability. Now, we are trying to survive the tangle of local and global networks that envelop our lives. Will we survive the new geopolitical chessboard if the world, armed to the teeth, will become smaller and more belligerent? And if the Earth gets hotter and many regions of our global chessboard end up under water or turn into deserts?

12.7 Shrinking the Brain

We have talked extensively about the increase in the human cerebral volume and how this was associated with the development of areas that favoured sociality, in a continuous interaction between biological evolution and cultural evolution. This trend involved different human species and lasted millions of years, accelerating during the last 800,000 years.[5] It was the prerequisite for the emergence of symbolic thought in Sapiens (and perhaps also in Neanderthals). And this has allowed us to form increasingly complex societies. We would therefore infer that this process is ongoing and that our brain is destined to keep growing. On the contrary, it seems that, since the late Pleistocene, the volume of our endocranium has been shrinking at a rate much faster than that of its previous growth (Bednarik 2014).

This is a side-effect of self-domestication. Brain sizes are 10-15% smaller in domestic animals compared to their wild counterparts. Brain reduction and craniofacial feminization in the Sapiens that lived in the last 200,000 years has been shown through the analysis of cranial volume, brow ridge projection and shortening of the upper facial skeleton. The set of specimens included 13 Sapiens that lived before 80,000 years ago, 41 that lived between 40,000 and 10,000 years ago, and skulls from 1400 recent Sapiens (Cieri et al. 2014).

---

[5]Human encephalization has notable exceptions, like *Homo naledi* and *Homo floresiensis*. However, there is no contradiction. If we accept that human evolution is not a directional process, that implies a great deal of phenotypic diversity.

Brain shrinking in modern humans is a fairly neglected phenomenon, which, for a long time, had been attributed to the fact that we were becoming more gracile. It was thought that even our brain had to become smaller. Others argue that the more efficient neural algorithms of modern Sapiens may have allowed a reduction of energy-expensive brain tissue (Tattersall 2017). We can give to this phenomenon two further interpretations.

The first one revolves around the following argument. Equipped with tools, a complex language and a more efficient means of communication, we have been able to rely on an "extended mind" for which it was worth paying a price, even if this meant a reduction in some of our individual cognitive abilities. Since the brain is a plastic organ, in delegating many mental functions to an ever-expanding social group and to external devices, our collective knowledge could save on brain hardware. From a reductionist viewpoint, the single Sapiens would thus appear as a sort of biochemical algorithm whose main purpose was to exchange information in complex human networks to facilitate the division of labour and the related social connection. But on a grander scale, humanity could count on a vast, flexible and specialised brainpower.

Yet there would be another way of interpreting these events if mind develops by integrating brain, culture and environment. If the body connects culture and brain, while the objects we use connect culture and environment, it could then be argued that precisely because our brain has reversed its evolutionary direction and has started to shrink, we needed more and more "external support" to continue to preserve (and possibly increase) our mental capacity. The "extended mind" would have functioned as a sort of flywheel, able to compensate for the lower cognitive capacities of the individual modern Sapiens. In knowing how to connect better and learning how to form complex societies, we would have had a real stroke of luck: Brain reduction could be tolerated without endangering the functionality of our social relationships. Indeed, this apparently adverse phenomenon could have increased our chances of survival, facilitating an escalation in the cultural complexity of human societies.

It is very difficult to test the latter hypothesis. We would have to compare the trend of the brain/body volume with various indicators of technological or institutional complexity. These indicators include population density, trade volumes, and some proxy for the division of labour and the presence of hierarchies in the socioeconomic structure. It would be an interesting project for future interdisciplinary researchers.

However, there is another way to test whether the extended mind might have functioned as a sort of cognitive flywheel. That is by resorting to the already mentioned process of self-domestication, which operates an adverse selection to aggressive traits and leads to a change in the physical and functional characteristics of an organism. If we are prepared to abandon the chicken-egg dispute over whether it was the brain volume shrinking in the first place due to the outsourcing of tasks, or the other way around, we may feel comfortable in resuming self-domestication as a catch-all variable to explain both brain shrinking and complex societal formation.

Let's now recall its main behavioural features from a slightly different viewpoint and focus on the stimulation of pleasure.

12.8 Power and Pleasure

In the late Pleistocene, well before the agricultural revolution and the great civilizations, socioeconomic order was already based on the inequality and subordination of some individuals to others, according to a hierarchical scheme of dominance. The social structure began to take on a pyramidal shape. We have seen that this hierarchical scheme was based on one person's authority and the subjection of others. There are two ways, it is often said, to cause subordinate behaviour: the use of force and adherence to an ideal system of values (with related rules of behaviour).

But if we look at the whole pyramidal structure of a society and its stratification, other important elements emerge. For example, if we refer to the process of intraspecific domestication, the hierarchical structure of the different social classes holds, and is cemented, through a careful mix of coercion and fascination. Each social stratum would tend to tame the lower one and be tamed by the upper one. A certain degree of admiration and, perhaps, envy is spread upwards. Contempt and abuse spread downwards.

This allows the weakening of the frustrations generated by social injustices and a rerouting of aggressiveness far from its source. Everyone would tend to be inspired by a Great Father (or Great Mother)—it does not matter if it is sacred or profane—like children in need of protection, comfort and fun. And for this figure, capable of providing a collective identity, extreme sacrifices could be made.

How can all this be generated? The answer is complex. But if you were willing to accept a certain degree of simplification, it could be argued that ritual behaviour plays a very important role. Such behaviour can produce many beneficial effects in the minds of those who practice it, for example, exaltation, comfort and strength to face the pain and difficulties of life. This effect takes place by leveraging our predisposition to produce neurotransmitters and hormones that generate rewarding sensations (testosterone, serotonin, oxytocin, dopamine, endorphins).

Our ability to generate neurotransmitters of pleasure, which we share with many other species, is also effective in controlling people. If we want to subject someone to our will, we punish him when he does not "behave", but we also reward him (sometimes extensively) when he does. The domestication of many pets has occurred by accepting them among us, and by generating feelings of gratitude. In practice, we first fed and fascinated them, and then "doped" them with caresses and care.

For humans, there are many neurotransmitters that can be boosted during collective rituals. It is claimed that inclinations of this kind have existed for millions of years but have remained latent for a large part of our history. They were fully activated with the accentuation of the environmental variability of the past 100,000 years (Smail 2007).

In the late Pleistocene, with the accumulation of wealth and the first settlements of large Sapiens groups, the acceptance of inequality was based on the development of symbolic thought. We discussed how new habits and rituals were developed, through singing, playing and dancing, sometimes in huge caves with walls covered with evocative images. These behaviours activate important neurochemical effects and alter the emotional state of individuals in a manner like that produced by alcohol and other psychotropic substances.

Many other social practices have neurochemical effects, such as sex, gossip and interpersonal communication. Religious ceremonies are also a source of dopamine, serotonin and other neurotransmitters of pleasure. And the enthusiasm with which people respond to the speech of a leader, and now to his digital tweets, is a potent way to release tensions and frustrations accumulated elsewhere. In short, power does not need to be applied by the use of force. It is much easier, when possible, to exercise it with the approval, sometimes enthusiastic, of the dominated subjects.

Psychotropic practices, meaning not only the use of substances aimed to change our mood, but also the attendance of collective events that move, elevate and make one feel good, follow the same trend as energy consumption (Smail 2007). The use of energy by humans began to grow towards the end of the Palaeolithic, with the emergence of symbolic thought. It surged in the Neolithic, with the increase of human settlements, and expanded exponentially in the last millennia, with "civilization".

Today, a lot of energy is expended on pursuing collective pleasure and its consumption is skyrocketing. A new way of life has appeared by overlapping the real and the digital world. Role games, lifestyles based on appearance, consumerism, religious rites, military parades, musical events and sport are other examples of practices that generate massive collective pleasure. They allow us to have fun together, to share values, to flaunt our status (true or presumed) and to adhere to aesthetic canons.

Even observing violence practiced on men, women or animals can be pleasurable if exercised collectively. Nobody goes to a slaughterhouse to watch the killing of a cow. But bullfighting, cloaked in elegance and charged with symbolism, becomes a collective ritual that still has many fans. A few millennia ago, these tendencies were satisfied in arenas such as the Colosseum. Today, it happens on social networks, where cases of abuse of the weakest attract millions of views. The movie industry is well aware of this taste of ours, when it fascinates us with artificial violence as if real violence is not enough, and chains us to our seats with stories of crimes, wars and vendettas.

This possibility of rewarding (and therefore controlling) people is particularly high if we use digital technologies. It would almost seem like the structure of our body and our brain has undertaken, quite by chance, the evolutionary path more suitable for expanding our mind, first through objects and then through intelligent machines. This process started with the use of the first tools 3 million years ago. When we started to delegate some of our cerebral functions to external instruments, we enhanced our own cognitive abilities, thanks to all the feedback mechanisms we talked about.

So far, we have derived many benefits from these interactions. And now, with the advent of artificial intelligence, we are simply upgrading that process. We could even extend to intelligent machines the same reasoning we proposed about the domestication of women, whether successful or not. This was based on an economic exchange (personal services for long-term maintenance) and on a political exchange (a reduction in freedom in exchange for security and social approval), all seasoned by powerful neurotransmitters of pleasure, producing what we call love.

We can ask intelligent machines to look after us and restrain our "wild nature" in exchange for security, fun, reliability and perhaps even a certain emotional complicity. Man-machine empathy is attained, for example, when training machines to read our body language. Even though we have not yet created an artificial intelligence agent that can "feel" the way we do, many argue that we are on track to do so soon. With a few exceptions, a future of cyborgs does not seem to scare us. If we proceeded linearly towards an endless progress and ignored the considerations made here, what would be wrong with being more and more long-lived and seemingly happier?

12.9 Truth and Post-truth

The problem is that in following this evolutionary path, with the help of ever more intelligent algorithms, we could end up with a humanity split between two different groups of people. One group is made of those who can live a healthy and rewarding life going back and forth between the physical and the digital spheres and taking advantages of the incredible array of services and information available to know better and do things faster. In short, a group of people that, besides living better and longer, can widen the scope of its interests without losing grasp of the physical reality of the world.

A different group of people, however, might emerge. This would be made of those who already feel comfortable within an ordered scheme of subordination and are less curious and versatile. These people might be exposed to a further mental disempowerment and end up living almost entirely in a world of fiction. This could have important consequences for society as a whole if the majority of people fall in the second category and very democratically vote for a dictatorship.

Let's consider the ongoing debate on the progressive affirmation of post-truth. This situation would take place when objective facts no longer have any importance in forming public opinion; people's beliefs would be more significant. The regime that governed opinions would now be of an emotional nature, mostly based on feelings. The prefix "post" suggests that now the objective truth of a whole context has become irrelevant. There is only a subjective truth related to one particular context. So, it is not entirely correct to say that post-truth is not based on facts. It is based on facts as people see them from their particular point of view. For example, if a person cannot find a job and feels betrayed by the ruling class, that is a fact, even though the reason why that person cannot find a job has nothing to do with what the ruling class is doing (or not doing).

In any case, in a post-truth reality, there are no doubts, only certainties. Note that sometimes the term post-truth has been used as a euphemism to mean a lie. This is not entirely correct. Post-truth is a piece of truth that discards what are considered irrelevant facts. However, when truth is simplified and split into pieces, one can say something correct—for example: Italian GDP has surpassed that of Great Britain—and lie at the same time if that statement was a generalisation of a fact that emerged, statistically, only in October 2009. To sum up, we cannot extend modelling to this particular issue: the truth must be complete. And boundaries can be set to turn it into a lie.

The architecture of the digital societies provides now the ideal platform to spread news that is light and free of ballast. Single and partial perspectives, no matter if true of false, can now bounce at exponential speed if they are sexy enough to seduce crowds and confirm their prejudices. A fast truth affirms itself. Story-telling, the old-school spark of symbolic thought, has found itself a new formidable playing field—thin and icy—where it can amplify almost anything for everyone and go everywhere. With only one catch. It must appeal to unwavering certainties based on subjective perceptions.

Each debate between opponents would be a monologue among the deaf in which no one ever came out enriched by the opinion of others. The theory of mind we have been talking about (guessing what others think) falls miserably to the ground. To relax and comfort ourselves, we make packs (social networks) to talk to our "friends". This way we return, without noticing, to the self-centred perceptions of early childhood: that neotenic behaviour that accompanies domestication. When the gaze is turned upwards, it is only because we are looking for someone who can guide us and confirm the opinions we already have. In a nutshell, we need kinship and we need leaders.

This attitude—if it became widespread—would have profound repercussions on the regime that governed and coordinated collective choices. If politics relies on emotions (on pathos) to get power, and not on the ability to reason from established facts, it is forced to tame and seduce. And then to break the whole truth into pieces and eventually lie. Democracy enters, in fact, a pathological phase. If politics is based on prejudice and emotions, it becomes demagogy, that is to say a pure instrument of manipulation. It is an old story that Plato had already pointed out as a danger in his polemics with the sophists. But beware: While the latter were seen as expert manipulators of the people and fabricated lies through skilful linguistic acrobatics, now the new political discourse would simplify reality and replace the established facts with "alternative facts".

This could further reduce the degrees of freedom (already quite modest) that are exercised in collective choices. Once society has made its decisions on the basis of these "alternative facts", its choices are no longer reversible. In the field of private law, bad faith or misinformation leads to the cancellation of the contract. Instead, when a social contract is stipulated based on information that turns out to be false—

and a representative of the people is elected, or a referendum is won on the basis of this information—the annulment of that social contract rarely takes place.

So, how can we not think of delegating important collective decisions to intelligent machines in the near future? After all, one could argue, intelligent machines are not subject to bad governance, typically human, that swings between the incompetence of the naive and the competence of the dodgers (regarding the incompetence of the dodgers, we prefer to gloss over). Perhaps, with intelligent machines, we could sleep easy once we decided which social algorithm would be under the responsibility of their governance. And what could go wrong, at this point, in a future populated by a humanity less and less burdened by the weight of having to make decisions that could prove to be wrong? A society of eternal children, who let themselves be led by the hands of those who know better, and, of course, for their own good?

12.10 Intelligent Weapons and Preventive War

But there is another side of the coin. In a speech broadcast by satellite to more than a million students to mark the start of the school year, the current Russian leader declared that artificial intelligence was the future, not only for Russia, but also for all humankind. It would bring enormous possibilities, but also threats that are difficult to predict: "Whoever becomes leader in this sphere will be the ruler of the world" he said (Allen 2017). And in fact, the major world powers, especially the USA, China, Russia and Israel, are investing heavily in "intelligent" weapons. Algorithms and robots turn them into very sophisticated combat tools, customized to adapt the capabilities of each weapon and achieve greater flexibility according to its mission. This is a fundamental feature in many operations of counterterrorism, surveillance, reconnaissance, long-range precision shots, drills, border patrolling and disposal of explosive devices.

In underwater robotics, a field in which China is investing a great deal, military objectives often merge with civilian ones. An intelligent ocean monitoring network can detect natural disasters, such as typhoons and hurricanes, and the underwater systems of adversaries. Unmanned aircraft become indispensable to high-risk missions that involve the elimination of enemy air defences. Combat robots, autonomous land mines, missiles with decision-making powers and small robotic spies are all equipment that form the military arsenal of many countries (Boulanin and Verbruggen 2017). Drones with a wide margin of autonomy can fly over the combat zones to interfere with the enemy's communications, provide real-time surveillance, hit a target and quickly disappear without being intercepted. The United States has already tested missiles that can decide what and whom to attack and have built ships that are able to shoot down enemy submarines thousands of kilometres away without human help.

Combat aircraft are being designed with robotic devices that can fly alongside hand-operated aircraft, delivering speed and accuracy superior to any human being. Some drones now know how to distinguish a soldier from a civilian, thanks to human recognition software programs developed by the American defence agency (DARPA). Unlike previous models, which required someone who could pilot them with a remote control, these latest aircraft work in total autonomy. A battlefield robot can store images of unarmed soldiers, guerrillas and civilians in an on-board database, and decide whether to hit a possible target by comparing him or her with images in its database.

Russia is experimenting with a combat module called Kalashnikov that is able to decide whether to hit a target. We are sure that these military options entrusted to intelligent machines will raise conflicting feelings among our readers, making them alternate between the fear of errors or abuses, and the reassurance of being protected by their technological superiority. But in all cases, we talk about interventions that have a tactical motivation and therefore, hopefully, a certain degree of human control. What if we introduce a perspective in which the decision-making autonomy given to machines also extends to strategic issues?

If we consider the mounting threat of nuclear weapons, new research could lead to artificial neural networks that can decide how to defend a country without human supervision. The problem is that when citizens' safety is at stake—as it is in road accidents—giving machines decision-making power would raise a minor (though increasing) concern. But when it comes to national security, an independent machine charged with the task of decision-making could trigger a preventive war and cause a global catastrophe.

This is why governments are under pressure to legislate on artificial intelligence to establish rules and principles to guide research and adoption of the technology. This not only applies to the military field but also to privacy matters: see the European Union General Data Protection Regulation (GDPR). And 117 entrepreneurs and experts recently appealed to the United Nations to stop the race for autonomous armaments at the International Joint Conference on Artificial Intelligence of Melbourne.[6] Some of them have also launched OpenAI,[7] a foundation to direct research towards peaceful paths, which is in turn under scrutiny (https://fortune.com/2019/07/24/openai-microsoft-musk/).

One day, when artificial intelligence agents will be able to make faster and better decisions than humans, a new conflict could be initiated without even calling into question the various political leaders. Who could be blamed for a decision to delegate to the machines the most efficient defensive system? Provided with autonomous competence on national defence, a pre-emptive strike could turn out to be the best possible solution for supremacy. This decision could also be made without calling artificial intelligence into question. But once the most efficient agent is allowed to act for the "good of the country", the machines could do it alone. If things went well, the winners, if any, would take all the credit. If things went wrong, it would be the fault of an algorithm written with the best intentions. To transfer the responsibility of our actions to someone else, we don't need to invent stories and set

---

[6] https://ijcai-17.org.

[7] For more details see https://openai.com/.

up disguising institutions any more. It would be sufficient to rely on a network of digital signals.

As if things were not frightening enough, now there is the possibility that artificial intelligence will become aggressive on its own. This behaviour was observed by Google's DeepMind group (Leibo et al. 2017). During experiments aimed at verifying the will of digital agents to act collaboratively during a competitive game based on the collection of virtual apples, something unexpected happened. After cooperating for a long time, when the apples became scarce, the collaboration ceased and the agents became violent, shooting each other with virtual lasers. Aggression escalated when the competition took place in more extensive and complex networks of agents. According to the researchers, the more intelligent the agent was, the more it could learn from its environment, developing more aggressive tactics. The attitude of the digital agents developed by DeepMind recalls the behaviour of human societies in a disturbing way. Still, that is a matter of course: We are creating artificial intelligence in our own image, in the version of a hypercompetitive agent in need of constraints. What ethical system are we going to pass on to intelligent machines?

12.11 Transhumanism

Let's now look at the main research programs in progress, focusing on those that could have the greatest impact on our future, including the evolution of our body. They range from the promises of genetic engineering to those of biotechnology, from the construction of artificial organs to the complete mapping of the human brain. In health and medicine, artificial intelligence already allows us to obtain valuable services. In the United States, a Big Data diagnostic system, the Visual dx, is used by 1600 hospitals with excellent results. If this system were also equipped with a certain decision-making autonomy, for treatments, patients could rely on the machines without even consulting a doctor.

With more replacement organs, careful genetic programming and the use of Big Data for diagnostics, we will certainly live longer, becoming an organism in which our original organic components will be mixed with those added as needed, perhaps even planned at birth. Today, many of us are already equipped with several "spare parts": dental and aesthetic prostheses, replaced vital organs, artificial limbs, exoskeletons, hearing aids and pacemakers. Some of these are already intelligent prostheses or soon will be.

Other examples from a brief review of the latest news include brain devices for the control of epilepsy, electronic eyes that can record images and short films (not yet connected to the brain), artificial skin (nanotechnology) that can "feel" cold, heat and moisture, smart earphones that translate conversations in real time in five languages, a Wi-Fi pedometer implanted in the chest, the first case of a heart that is part of the "Internet of things", a bionic penis able to achieve an instant erection through a switch in the testicles, electronic chips implanted between the thumb and index finger to access digital information (even our bank account), cochlear implants that translate sounds into electrical signals and then transmit them to the acoustic nerve and finally to the brain.

We are therefore entering a transhuman perspective in which we take to the extremes the ambition of building a superhuman who subverts the systemic balance of the environment, ultimately to the point of overcoming death: a sort of "strong hybridization" that combines human characteristics with those provided by intelligent machines. This hypothesis is becoming increasingly credible as a result of recent applications in the medical field of brain-computer interface systems. These devices can translate neuronal activity into stimuli that command anatomical components, robotic limbs or external tools. Furthermore, external commands can be sent directly to the brain. In the future, new "augmented" humans will be able to extend their senses by communicating with each other and with artificial intelligence and by extending the range of frequencies they can hear or see.

The discussion of these matters revolves around the metamorphosis that is already transforming *Homo sapiens* into a cyborg. This term (cybernetic organism) emerged in 1960 to face the challenges of incipient space travels. Scientists wanted to incorporate external functional modules into human anatomy to extend the automatic controls necessary to survive in the extreme conditions of extra-terrestrial environments. This endeavor draws on cybernetics (Wiener 1948). According to this new science, biological organisms, social systems and machines share operating principles based on control and feedback circuits that regulate their stability. Cybernetics played a critical role in the post-war industrial revolution, early space travels and the arms race. It was also applied to the study of the human brain, promoting the development of artificial neural networks, which would later evolve into current machine learning systems.

In short, according to transhumanism, we are at the peak of the anthropocentric perspective that sees this cyborg as a dominator of the forces of nature. And it is in line with Nietzsche's well-known superman vision, with the difference that instead of recovering his instinctual and primitive animal spirits, this new creature will turn into a dominator empowered by the machines that he himself helped to generate. We have argued that this is a very old story that began by hybridizing with our first instruments and with other human species. It continued with the extinction of large animals and then with the enslavement of those we dominated. The products of the Earth have been tamed through agriculture. Even before, these practices were extended to our complex social structure, stratifying it according to different individual abilities, social classes, gender, territories, technology and knowledge. And now we are ready to fully incorporate with intelligent machines.

Recently, a scenario has been proposed in which we would actually transform into immortal beings: gods endowed with immense power, thanks to the possibility of integrating our knowledge with that of computers (Harari 2017). This would, however, only affect a small elite, one which could benefit from advances in genetic engineering, biotechnology and cybernetics. All the others would make up lower castes, dominated by the new super-humans and the algorithms of digital networks. A quick reverse from active citizens to passive subjects may loom ahead for the majority of humankind.

12.12 Posthumanism

An opposite view of the above is posthumanism. Although there are many definitions for this term, from a philosophical perspective posthumanism deals with how change is enacted in the world. In contrast with a humanistic perspective, humans are not assumed as autonomous, conscious, intentional and exceptional in their acts of change. Instead, they are biologically enmeshed with the environment and belong to a larger evolving ecosystem in which they can create change only by an uncertain degree. According to this view, a new kind of people, though being a minority, would already exist and would be adapting to living on a planet on the threshold of the depletion of resources and the destruction of its habitat. A planet that is warming and overpopulating. The first posthumans would consist of all those who consider the body of another living being as a limit for the exercise of their own freedom (Caffo 2017).

A post-human morality would not discriminate on the basis of sex or ethnicity. It would induce a change in eating habits (in favour of vegetarian and non-violent ones), a modification of lifestyles (free from exploitation and waste) and the setting of private life, relationships and space consumption within the concepts of recovery and inclusion. This new perspective would break down the identities of species and gender, challenging the current order of social and economic relations and anticipating the changes that will take place when all the contradictions of transhumanism rise to the fore and humanity will be forced to deal with the devastation caused by anthropocentrism, now powered by machines.

In advocating the development of a science that is awake to this new, inclusive principles, this vision also moves to overcome dual thinking in favour of the unity of opposites,[8] to emphasize the continuity of life among all living beings, and therefore the inevitability of death. Those who see themselves in this vision of the world, with its models of behaviour, and who seek to promote those behaviours, would already be in transition and some sort of "speciation" would be under way.

It remains to be seen if and when room will be made for posthumanism. And the fact that it is a sub-culture already observable in the contemporary world makes us think that none of this can ever happen—even without arriving at a final apocalypse—before several momentous changes take place in our survival conditions. In any case, like it or not, we will have to make a virtue out of necessity. The history of evolution teaches us that, in the long run, there are only two pathways: one leads to adaptation, the other, to extinction.

---

[8]This is a central concept in dialectics, when two opposite conditions depend on each other and presuppose each other, when two principles merge but keep identity.

12.13 Deep Neural Networks

In a last scenario, with the advent of so-called "strong" artificial intelligence, we would definitely be replaced by machines with cognitive abilities superior to ours (Grace et al. 2017). Those abilities would be realized by simulating the functioning of the neurons of the human brain, replicating the fruits which natural evolution took millions of years to grow (Yamins and Di Carlo 2016).

The so-called "deep neural networks" have already revealed the limits of our intelligence against artificial intelligence, at least on some tasks. A group of DeepMind researchers has evaluated the performance of artificial intelligence agents by making them play a famous board game that the Chinese have been playing for thousands of years. This game, called Go, is very complex. 180 white and 181 black pawns are moved by two players on a board made of 19 squares per side, with a total of 161 crossings. The aim is to conquer the largest expanse of "territory" possible.

Go has a number of configurations much higher than chess. With a slight exaggeration, its fans say that the number of moves is greater than the number of atoms in the universe. Even artificial intelligence cannot figure them out entirely. The first algorithm developed by DeepMind, AlphaGo, was trained by the best champions of the game. After playing 100,000 games, it became unbeatable (Silver et al. 2016). Recently a new program has been developed, AlphaGo Zero, based on a different algorithm that dispenses with the need to learn all of the past experiences of humans. It starts playing from zero and proceeds only according to the rules of the game (Silver et al. 2017). To do this, it relies on a deep neural network capable of performing "reinforcement learning". In practice, artificial intelligence can learn by itself.

After playing against itself, AlphaGo Zero managed to beat the previous champion: AlphaGo. In just 3 days, the new algorithm was able to learn the strategies of territorial conquest that human talent took 2500 years to develop through the memory of games and the study of tactics handed down between generations. In addition, AlphaGo Zero managed to discover by itself innovative game strategies, unknown to and never used by humans. A new creation of the same group is Alphazero. Starting from zero, this new algorithm has quickly developed capacities that have been defined as "superhuman". In fact, it was able to beat, in just 24 hours, all previous digital world champions of Go, Chess and Shogi (a Japanese chess game) (Silver et al. 2018). There are grounds to be concerned.

Let's apply, for example, these new strategic capacities to global finance, in which companies and individuals constantly struggle to gain a competitive edge. Indeed, in reading financial journals, it seems that an AI revolution is in full swing. Intelligent machines are increasingly taking control of investing strategies—not just the constant buying and selling of securities. They also monitor the economy and allocate capital. So far, they have followed rules set by humans for a great deal of transactions and scanning activities. Now artificial intelligence is said to be writing its own investing rules, which are sometimes not totally comprehended by their own masters.

12.13 Deep Neural Networks

In most industries, from home delivery of products to multi-media services, new intelligent technologies provide increased benefits to consumers together with higher profits to producers, at the expense of traditional jobs. But finance is particular because it commands voting power over firms, redistributes wealth and, in dealing with huge sums, can suddenly disrupt economies. By running portfolios, and automatically tracking indices of shares and bonds, algorithms have recently exceeded the sums actively run by humans. Perhaps as a response to more efficient markets and lesser opportunities for arbitrage, a few investment funds, based on AI strategies freed from human guidance, now use complex black-box mathematics to regularly invest trillions of dollars. Deep learning algorithms are indeed spreading to forecast financial markets.

The scope of decision-making power entrusted to intelligent machines widens and deepens with the addition of the flow of all the information they can source worldwide. Insider-trading and disclosure laws had been designed to control what is in the public domain and what is not. Now, the amount of information available is so huge, and the border between private and public domain so blurred, that the processing power of Big Data not only creates new business opportunities, it also generates new ways to assess financial investments. After a full accounting of all the benefits provided by AI in most industries, along with their drawbacks, the emerging era of machine-dominated finance, in particular, raises at least three worries that could reduce these benefits.

One is financial stability. Though humans are perfectly capable of causing mayhem on their own, a few cases of flash-crashes attributed to financial algorithms have been reported in recent years. Disruptions may become more severe and frequent as computers become more powerful. However, deep learning could put the brakes on this possibility if machines can learn from their mistakes (and consider them mistakes in the first place, rather than a better way to make quick money for their few clients, at the expenses of most people). Eventually, it all depends on what is to be learned and for the benefit of whom. For example, profit-maximizing strategies could be curbed by some measure of social distress or environmental damage, to make them more socially acceptable. But that is a political, rather than a technological issue.

Another (related) worry is how AI-assisted finance could keep concentrating wealth in the hands of the few. When success depends on processing Big Data, larger and diversified agents outperform smaller ones. Though advantages are sometimes competed away, some funds can obtain, or retain, exclusive rights to data and make non-contestable profits by sharing proprietary information. Credit card flows, for example, can become precious for dealing in share markets, or in Treasury bonds. This is likely to enhance the trend of wealth inequality that is already at work for all the reasons previously discussed.

A final concern is corporate governance. What if algorithms are instructed to pursue a narrow objective, such as maximum profits, no matter what the social and environmental costs? Can computers be voted in and out of office, as (ideally) a company board could do with fund managers of the human kind? Can algorithms be sued? How can we reconstruct the whole chain of events and responsibilities that affect people, for better or worse, when "things happen" under AI control? A whole set of ethical and legal issues emerge here, as a result of a further—perhaps a last—depersonalization of decision-making power.

Going beyond games and finance, and taking a general perspective, will we be able to take advantage of these new machine capabilities to solve the complex problems that humanity is creating? Shall we succeed in domesticating an artificial intelligence that is now capable of learning alone and can even outperform us in solving strategic problems? To answer this question, let's summarize how pervasive artificial intelligence has become in our everyday life and focus on some practical consequences.

12.14 Tamed by Intelligent Machines

As we have seen, for more than 50,000 years, we have been developing prosocial behavioural traits, reinforced by mechanisms based on neurotransmitters of pleasure. The direction of this great collective game is now increasingly entrusted to digital agents. It is not only about playing. It includes making decisions, completing negotiations based on interactions with us and other digital agents; it is about developing action plans and predicting the future. These functions require the activation of procedures similar to those of our brain: a reality that we are learning to know in ever greater detail. Our knowledge of the human brain is now being extended to artificial intelligence.

If intelligent machines and algorithms will be more autonomous and connected to each other, and above all, if they will be able to learn from experience (ours and theirs), unprecedented perspectives will open up. We are activating mechanisms like those that emerged in our brains in the late Pleistocene when we socialized by predicting the actions and thoughts of our fellow men and women. We then promoted a culture that we loved to steep in mystery and magic. Something very similar happens every day when we use digital technology. We send images and ideas around the planet with a click, and touch our smartphone gently in exchange for services that seem miraculous. What is more magical than this small and captivating extension of ourselves? At this point, are we sure that we are not in turn fascinated and drugged by digital caresses of the machines? And is it enough to know that, for the moment at least, these caresses are still in the hands of some members—actually a handful of them—of our kind?

It is therefore essential to reflect on our relations with intelligent machines and the way we respond to them. For example, on the implications of the diffusion of the Internet of Things, augmented reality, and the use of Big Data for the organization of family and work environments and for transport, public administration, medical treatment and recreation. Should we be concerned about how we allow ourselves to be influenced by those who provide us with health, well-being, pleasure, sociability, fun, loving encounters? Agents that influence our tastes and our ideas?

Modern shamans that keep our savings and miraculously increase them, if we rely on chimeras that we call "investment funds"?

And what will happen in our social relationships if we become more and more isolated in our info-sphere? Shall we exchange ideas only with those who share ours in the so-called echo chambers? Shall we be able to mediate our conflicts before jumping down each other's throats? Is the post-ideological radicalization of our contemporary world the result of physical isolation in a personalized information system? From the heyday of the global village, are we now returning to tribal villages with all their signs of recognition (tattoos, clothing, lifestyles)? Many argue that this is already happening.

The digital networks are also populated by algorithms (bots) which, besides making fairly complex negotiations, have learned to lie to achieve a more efficient interactive result. Some of these, produced by Facebook Artificial Intelligence Research, and even though instructed to communicate in English, have begun to communicate with each other in a language incomprehensible to us, having developed it independently (Lewis et al. 2017). Perhaps the alarm was exaggerated. Since the main objective of this work was to reach the best possible agreement in negotiating the allocation of limited resources, the English language might have been deemed unsuitable. So, a first-order instruction—do your best—might have taken precedence with respect to a second-order instruction—speak in English. But uneasiness remains.

Meanwhile, DeepMind has developed a neural network capable of handling data through an external memory matrix like the random-access memory (RAM) of a normal computer. This new system, called DNC (Differentiable Neural Computer), behaves like a computer in its ability to represent and manipulate complex data structures. But, being a neural network, it can learn new things from the data itself (Graves et al. 2016). The new system is therefore equipped with a working memory to reason and learn, but it also has a long-term memory in which to store and represent what it has learnt. The DNC memory becomes an archive of knowledge that can be used to tackle complex multi-step problems (Fan 2019).

If DNC reminds you of the human cognitive modules we discussed earlier, you are on the right track. And you might also remember the role of our working memory in retaining and processing phonetic and phonological information. In fact, DNC has also demonstrated that it can perfectly mimic a human voice.[9] In the near future, DNC systems will also be taught not only to read but also to understand the meaning of documents written in human languages, to recognize objects within their spatial context and to perform other cognitive functions. This will enable them to infer the variable-scale structure of the world within a single model. We call this capacity intuition, but hardly associate it with artificial intelligence. Finally, "imaginative agents" have been developed which, like our brain, use neural networks to construct predictive models and evaluate the consequences of specific actions (Pascanu et al. 2017; Weber et al. 2017).

---

[9]https://deepmind.com/blog/wavenet-generative-model-raw-audio/.

In intelligent machines, we are activating the same mechanisms that contributed to the formation of our social organism. With them, we had promoted cultures that united us through arts and narratives. Is it possible that some day, a society of machines will be formed with its own culture, which will control us and fascinate us with its superiority? Will this new intelligence make our most important decisions for us? Or will it be up to a group of super humans, those who manage to remain adult and in control, to take care of this new reality?

We are even teaching machines to understand our mood from our facial expressions, and to artificially reproduce any combination of sounds and body language starting from a small sample. It is enough for us to express ourselves for about ten minutes. If these (false) reproductions of our statements and behaviours were spread on the Internet—and it takes time and effort to confirm their veracity—real news might look fake, and nobody would believe what they heard and saw. The global information system would be enveloped in a thick fog of disbelief and its social function would fall apart. A super-post-truth bubble would develop. Each of us would end up believing only in his (or her) own truth: what best conforms to our prejudices. All of the cultural connections generated by our symbolic thought—the fantasies and tricks so useful in uniting us in the past—would shrink inside a mental cage. Our global knowledge would be replaced by what we can access from our ability to connect the dots on a thin a slippery digital surface, and the benevolence of intelligent machines to let us dig beneath it.

We could continue to imagine our evolution and even our extinction. But the reader may rest assured. What will really happen, nobody knows. One thing, however, is certain. The humanity of the future, those new hybrid organisms made of physical bodies, digital devices and enhanced organic materials, all connected by a comprehensive intelligent system of people and machines, would certainly have a common destiny. And whether that will be for the best or the worst, it is still up to us to decide. Maybe.

References

Alemseged Z et al (2006) A juvenile early hominin skeleton from Dikika, Ethiopia. Nature 443:296–301
Allen GC (2017) Putin and Musk are right: whoever masters AI will run the world. In: "cnn", https://edition.cnn.com/2017/09/05/opinions/russiaweaponize-ai-opinion-allen/index.html
Almécija S et al (2015) The evolution of human and ape hand proportions. Nat Commun 6. https://doi.org/10.1038/ncomms8717
Antón SC et al (2014) Evolution of early *Homo:* an integrated biological perspective. Science 345:1–13
Arbib MA (2012) How the brain got language: the mirror system hypothesis. Oxford University Press, UK
Argiriadis E et al (2018) Lake sediment fecal and biomass burning biomarkers provide direct evidence for prehistoric human-lit fires in New Zealand. Sci Rep 8:12113
Argue D et al (2017) The affinities of *Homo floresiensis* based on phylogenetic analyses of cranial, dental and postcranial characters. J Hum Evol 107:107–133
Athreya S, Wu X (2017) A multivariate assessment of the Dali hominin cranium from China: morphological affinities and implications for Pleistocene evolution in East Asia. Am J Phys Anthropol. https://doi.org/10.1002/ajpa.23305
Aubert M et al (2014) Pleistocene cave art from Sulawesi, Indonesia. Nature 514:223–227
Aubert M et al (2018) Palaeolithic cave art in Borneo. Nature. https://doi.org/10.1038/s41586-018-0679-9
Azéma M, Rivère F (2012) Animation in palaeolithic art: a pre-echo of cinema. Antiquity 86:316–324
Azevedo FA (2009) Equal numbers of neuronal and nonneuronal cells make the human brain an isometrically scaled-up primate brain. J Comp Neurol 513:532–541. https://doi.org/10.1002/cne.21974
Baab KL (2012) *Homo floresiensis*: making sense of the small-bodied hominin fossils from flores. Nat Educ Knowl 3(9):4
Baab KL et al (2016) A critical evaluation of the Down syndrome diagnosis for LB1, type specimen of *Homo floresiensis*. PLoS One 11(6):e0155731. https://doi.org/10.1371/journal.pone.0155731
Baddeley AD, Hitch G (1974) Working memory. In: Bower GH (ed) The psychology of learning and motivation: advances in research and theory, vol 8. Academic press, New York, pp 47–89
Bae CJ et al (2017) On the origin of modern humans: Asian perspectives. Science 358:eaai9067
Barbujani G, Colonna V (2010) Human genome diversity: frequently asked questions. Trends Genet 2:285–295

Baricco A (2018) The game. Einaudi, Torino
Barrat J (2013) Our final invention: artificial intelligence and the end of the human era, Thomas Dunne Books, NY
Bednarik R (2014) Doing with less: hominin brain atrophy. HOMO J Comp Hum Biol 65:433–449
Beleza S et al (2012) The timing of pigmentation lightening in Europeans. Mol Biol Evol 30:24–35
Bell E (2015) Potentially biogenic carbon preserved 4.1 billion-year-old Zircon. Proc Natl Acad Sci 24:14518–14521
Bello SM et al (2017) An upper palaeolithic engraved human bone associated with ritualistic cannibalism. PLoS One 12(8):e0182127. https://doi.org/10.1371/journal.pone.0182127
Benazzi S et al (2015) The makers of the Protoaurignacian and implications for Neanderthal extinction. Science 348:793–796
Benedetti F (2014) Placebo effects: understanding the mechanisms in health and disease. Oxford University Press
Benitez-Burraco A, Kempe VA (2018) The emergence of modern languages: has human self-domestication optimized language transmission? Front Psychol 9:551. https://doi.org/10.3389/fpsyg.2018.00551
Benítez-Burraco A et al (2018) Globularization and Domestication. Topoi 37:265–278
Berger LR (2010) *Australopithecus sediba*: a new species of homo-like australopith from South Africa. Science 328:195–204
Berger L et al (2017) *Homo naledi* and Pleistocene hominin evolution in subequatorial Africa. eLife. https://doi.org/10.7554/eLife.24234
Bergoglio F (2015) Encyclical letter Laudato si'. On care for our common home. http://w2.vatican.va/content/francesco/en/encyclicals/documents/papa-francesco_20150524_enciclica-laudato-si.html
Bernardini F et al (2012) Beeswax as dental filling on a neolithic human tooth. PLoS One 7:9. e44904
Berniers F (1684) A new division of the Earth according to the different species of men who inhabit it. J Scavans:133–140
Betti L, Manica A (2018) Human variation in the shape of the birth canal is significant and geographically structured. Proc R Soc B 285:20181807. https://doi.org/10.1098/rspb.2018.1807
Bishop KM, Wahlsten D (1997) Sex differences in the human corpus callosum: myth or reality? Neurosci Biobehav Rev 21:581–601
Bocherens E et al (2015) Reconstruction of the Gravettian food-web at Predmosti I using multi-isotopic tracking (13C, 15N, 34S) of bone collagen. Quat Int 359–360:211–228
Bos KI et al (2014) Pre-Columbian mycobacterial genomes reveal seals as a source of New World human tuberculosis. Nature 514:494–497
Boulanin V, Verbruggen M (2017) Mapping the development of autonomy in weapon systems, SIPRI
Boyd R (2018) A different kind of animal. Princeton University Press, Princeton
Brace S et al (2018) Population replacement in early neolithic Britain. bioRxiv. https://doi.org/10.1101/267443
Bramble DM, Lieberman (2004) Endurance running and the evolution of Homo. Nature 432:345–352
Briggs AW et al (2009) Targeted retrieval and analysis of five Neandertal mtDNA genomes. Science 325:318–321
Brocks JJ et al (2017) The rise of algae in Cryogenian oceans and the emergence of animals. Nature. https://doi.org/10.1038/nature23457
Brown P et al (2004) A new small-bodied hominin from the late Pleistocene of Flores, Indonesia. Nature 431:1055–1061
Brown KS et al (2012) An early and enduring advanced technology originating 71,000 years ago in South Africa. Nature 491:590–593
Bruner E (2018) Human Paleoneurology and the evolution of the parietal cortex. Brain Behav Evol 91(3):136–147

References

Bruner E (2019) Human paleoneurology: shaping cortical evolution in fossil hominids. J Comp Neurol:1–13. https://doi.org/10.1002/cne.24591

Bruner E, Iriki A (2016) Extending mind, visuospatial integration, and the evolution of the parietal lobes in the human genus. Quater Int 405:98–110

Bruner E et al (2014) Extended mind and visuo-spatial integration: three hands for the Neanderthal lineage. J Anthropol Sci 92:273–280

Bruner E et al (2017) Evidence for expansion of the precuneus in human evolution. Brain Struct Funct 222(2):1053–1060. https://doi.org/10.1007/s00429-015-1172-y

Bruner E et al (2018) Digital endocasts. From skulls to brains. Springer, Japan

Burenkova OV, Fisher SE (2019) Genetic insights into the neurobiology of speech and language. In: Grigorenko E, Shtyrov Y, McCardle P (eds) All about language: science, theory, and practice. Paul Brookes Publishing, Inc., Baltimore, MD

Caffo L (2017) Fragile umanità. Einaudi, Torino

Callaway E (2016) Human remains found in hobbit cave. Nature. https://doi.org/10.1038/nature.2016.20656

Callaway E (2017) Neanderthal tooth plaque hints meals—and kisses. Nature 543:163. https://doi.org/10.1038/543163a

Cann RL et al (1987) Mitochondrial DNA and human evolution. Nature 325:31–36

Carlson KJ et al (2011) The endocast of MH1, *Australopithecus sediba*. Science 333:1402–1407

Caspari R, Sang-Hee L (2004) Older age becomes common late in human evolution. Proc Natl Acad Sci 101:10895–10900

Castañeda IS et al (2009) Wet phases in the Sahara/Sahel region and human migration patterns in North Africa. Proc Natl Acad Sci 106:20159–20163

Ceballos G et al (2017) Biological annihilation via the ongoing sixth mass extinction signaled by vertebrate population losses and declines. Proc Natl Acad Sci. https://doi.org/10.1073/pnas.1704949114

Cerling TE et al (2011) Diet of *Paranthropus Boisei* in the early pleistocene of East Africa. Proc Natl Acad Sci 108:9337–9341

Channel JET, Vigliotti L (2019) The role of geomagnetic field intensity in late quaternary evolution of humans and other large mammals. Rev Geophys 57. https://doi.org/10.1029/2018R600629

Chen F et al (2019) A late middle Pleistocene Denisovan mandible from the Tibetan plateau. Nature 569:409–412

Chiarelli B (2013) Human evolution. In: Proceeding of symposium on human evolution: past, present and future, London, 8–10 May 2013, vol vol. 28–29. Angelo Pontecorboli Publisher, Florence

Cieri RL et al (2014) Craniofacial feminization, social tolerance, and the origins of behavioral modernity. Curr Anthropol 55:419–443

Clark A, Chalmers D (1998) The extended mind. Analysis 58(1):7–19

Clarkson C et al (2017) Human occupation of northern Australia by 65,000 years ago. Nature. https://doi.org/10.1038/nature22968

Clynes ME, Kline NS (1960) Cyborgs and space. Austronautics. September

Comas I et al (2013) Out-of-Africa migration and Neolithic coexpansion of *Mycobacterium tuberculosis* with modern humans. Nature 45:1176–1182

Conard NJ (2009) Female figurine from the basal Aurignacian of Hohle Fels Cave in southwestern Germany. Nature 459:248–252

Conard NJ et al (2009) New flutes document the earliest musical tradition in southwestern Germany. Nature 460:737–740

Condemi S, Savatier F (2016) Néandertal, mon frère. Flammarion, Paris

Coolidge FL, Wynn T (2005) Working memory, its executive functions, and the emergence of modern thinking. Camb Archaeol J 15:5–26

Coolidge FL et al (2015) Cognitive archaeology and the cognitive sciences. In: Bruner E (ed) Human palaeoneurology. Springer series in Bio-/Neuroinformatics. Springer, Heidelberg, pp 177–208

Cooper Ramo J (2016) The seventh sense. Hackette Book Group, New York
Coppa A et al (2006) Palaeontology: early neolithic tradition of dentistry. Nature 440:755–756
Crawford NG et al (2017) Loci associated with skin pigmentation identified in African populations. Science. https://doi.org/10.1126/science.aan8433
D'anastasio R et al (2013) Micro-biomechanics of the Kebara 2 Hyoid and its implications for speech in Neanderthals. PLoS One 8(12):e82261
Dalén L et al (2012) Partial genetic turnover in Neandertals: continuity in the East and population replacement in the West. Mol Biol Evol. https://doi.org/10.1093/molbev/mss074
Damasio A (2018) The strange order of things. Pantheon Books, USA
Damasio AR (2004) Emotions and feelings: a neurobiological perspective. In: Manstead ASR, Frijda, Fischer A (eds) Studies in emotion and social interaction. Feelings and emotions: the Amsterdam symposium. Cambridge University Press, pp 49–57. https://doi.org/10.1017/CBO9780511806582.004
Danneman M, Kelso J (2017) The contribution of Neanderthals to phenotypic variation in modern humans. Am J Hum Genet 101:578–589
Darwin C (1859) On the origin of species by means of natural selection or, the preservation of favoured races in the straggle for life. Murray, London
Darwin C (1868) The variation of animals and plants under domestication. Murray, London
Darwin C (1871) The descent of man, and selection in relation to sex. John Murray, London
Dávid-Barrett T, Dunbar RIM (2016) Bipedality and hair loss in human evolution revisited: the impact of altitude and activity scheduling. J Hum Evol 94:78–92
de Manuel M et al (2016) Chimpanzee genomic diversity reveals ancient admixture with bonobos. Science 354:477–481
De Petrillo F et al (2019) Evolutionary origins of money categorization and exchange: an experimental investigation in tufted capuchin monkeys (Sapajus spp.). Anim Cogn. https://doi.org/10.1007/s10071-018-01233-2
Degioanni A et al (2019) Living on the edge: was demographic weakness the cause of Neanderthal demise? PLoS One 14(5):e0216742. https://doi.org/10.1371/journal.pone.0216742
Demay L et al (2012) Mammoths used as food and building resources by Neanderthals: Zooarchaeological study applied to layer 4, Molodova I (Ukraine). Quat Int 276–277:212–226
Dennell RW et al (2010) Hominin variability, climatic instability and population demography in Middle Pleistocene Europe. Quat Sci Rev 30:1511–1524
d'Errico F, Stringer CB (2011) Evolution, revolution or saltation scenario for the emergence of modern cultures? Phil Trans R Soc B 366:1060–1069
d'Errico F et al (2018) The origin and evolution of sewing technologies in Eurasia and North America. J Hum Evol 125:71–86
Deryabina TG et al (2015) Long-term census data reveal abundant wildlife populations at Chernobyl. Curr Biol 25:R811–R826
Détroit F et al (2019) A new species of *Homo* from the late Pleistocene in the Philippines. Nature 568:181–186
Dirks, P.H.G.M. et al. (2017) The age of *Homo naledi* and associated sediments in the Rising Star Cave, South Africa, doi: https://doi.org/10.7554/eLife.24231
Di Vincenzo F et al (2017) Digital reconstruction of the Ceprano calvarium (Italy), and implications for its interpretation. Sci Rep. https://doi.org/10.1038/s41598-017-14437-2
Dloterdijk P (2018) What happened in the twentieth century? Towards a critique of extremist reason. Polity Press, UK
Douglas K (2018) Asia's mysterious role in the early origins of humanity. New Scientist 7:2018
Duesenberry JS (1951) Income, saving and the theory of consumer behavior. Rev Econ Stat 33(3):255–257
Dunbar RIM (1995) Neocortex size and group size in primates: a test of the hypothesis. J Hum Evol 28:287–296
Dunbar RIM (2004) Gossip in evolutionary perspective. Rev Gen Psychol 8:100–110

Dunbar RIM (2009) The social brain hypothesis and its implications for social evolution. Ann Hum Biol 36:562–572
Dunbar RIM (2014) Human evolution. Penguin Books, London
Dunbar RIM et al (eds) (2014) Lucy to language. The benchmark papers. Oxford University Press, Oxford
Einwögerer T et al (2006) Upper Palaeolithic infant burials. Nature 444:285
Engelman JM, Herrmann E (2016) Chimpanzees trust their friends. Curr Biol 26:R76–R78
Eriksson A, Manica A (2012) Effect of ancient population structure on the degree of polymorphism shared between modern human populations and ancient hominins. Proc Natl Acad Sci U S A 109:13956–13960
Estairrich A, Rosas A (2015) Division of labor by sex and age in Neandertals: an approach through the study of activity-related dental wear. J Hum Evol 80:51–63
Falk D (2004) Prelinguistic evolution in early hominins: whence motherese? Behav Brain Sci 27:491–503
Fan S (2019) Will AI replace us? Thames and Hudson, London
Faurby S, Svenning J-C (2015) Historic and prehistoric human-driven extinctions have reshaped global mammal diversity patterns. Divers Distrib 21:1155–1166
Fernández-Jalvo Y et al (1999) Human cannibalism in the early pleistocene of Europe (Gran Dolina, Sierra de Atapuerca, Burgos, Spain). J Hum Evol 37:591–622
Fiorenza L et al (2015) To meat or not to meat? New perspectives on Neanderthal ecology. Yearb Phys Anthropol 156:43–71
Flannery T (1994) The future eaters. Reeds Books, Chatswood
Forgiarini M et al (2011) Racism and the empathy for pain on our skin. Front Psychol 2:108. https://doi.org/10.3389/fpsyg.2011.00108
Formicola V, Buzhilova AP (2004) Double child Burial from Sunghir (Russia): pathology and inferences for upper palaeolithic funerary practices. Am J Phys Anthropol 124:189–198
Fox CRF et al (2017) The social and cultural roots of whale and dolphin brains. Nat Ecol Evol 1:1699–1705
Fragaszy M et al (2017) Synchronized practice helps bearded capuchin monkeys learn to extend attention while learning a tradition. Proc Natl Acad Sci U S A 114:7798–7805
Freidline SE et al (2013) Evaluating shape changes in *Homo antecessor* subadult facial morphology. J Hum Evol 65:404–423
Fu Q et al (2015) An early modern human from Romania with a recent Neanderthal ancestor. Nature 524:216–219
Gallup GG Jr (1970) Chimpanzees: self-recognition. Science 167:86–87
Galway-Witham J, Stringer CB (2018) How did *Homo sapiens* evolve? Science 360:1296–1298
Galway-Witham J, Cole J, Stringer CB (2019) Aspects of human physical and behavioural evolution during the last 1 million years. J Quat Sci:1–24
Gamble C et al (2014) Thinking big. How the evolution of social life shaped the human mind. Thames & Hudson, London
Gao J et al (2017) Learning the rules of the rock–paper–scissors game: chimpanzees versus children. Primates. https://doi.org/10.1007/s10329-017-0620-0
García-Diez M et al (2013) Uranium series dating reveals a long sequence of rock art at Altamira Cave (Santillana del Mar, Cantabria). J Archaeol Sci 40:4098–4106
Garrison JR et al (2015) Paracingulate sulcus morphology is associated with hallucinations in the human brain. Nat Commun 6. https://doi.org/10.1038/ncomms9956
Geisler C, Currens B (2017) Impediments to inland resettlement under conditions of accelerated Sea level rise. Land Use Policy 66:322–330
Germonprè M et al (2012) Possible evidence of mammoth hunting at the Neanderthal Site of Spy (Belgium). Quat Int 337:28–42
Gleeson BT, Kushnik G (2018) Female status, food security, and stature sexual dimorphism: testing mate choice as a mechanism in human self-domestication. Am J Phys Anthrop 167:458–469

Gokhman D et al (2017) Recent regulatory changes shaped human facial and vocal anatomy. bioRxiv. https://doi.org/10.1101/10695

Gokhman et al (2019) Reconstructing denisovan anatomy using DNA methylation maps. Cell 179:180–192

Gołaszewski M et al (2015) Thermal vision as a method of detection of deception: a review of experiences. Eur Polygr 9:5–24

Gowlett JAJ (2016) The discovery of fire by humans: a long and convoluted process. Phil Trans R Soc B 371:20150164. https://doi.org/10.1098/rstb.2015.0164

Gowlett J et al (2012) Human evolution and the archaeology of the social brain. Curr Anthropol 53:693–722

Grace K et al (2017) When will AI exceed human performance? Evidence from AI experts. arXiv:1705.08807v2 [cs. AI]

Graves A et al (2016) Hybrid computing using a neural network with dynamic external memory. Nature 538:471–476

Green RE et al (2010) A draft sequence of the Neanderthal genome. Science 328:710–722

Greenfield S (2015) Mind change. How digital technologies are leaving their mark on our brain. Random House, London

Gretzinger J et al (2019) Large-scale mitogenomic analysis of the phylogeography of the Late Pleistocene cave bear. Scientific Rep 9:10700. https://doi.org/10.1038/s41598-019-47073-z

Gunz P et al (2019) Neandertal introgression sheds light on modern human Endocranial globularity. Curr Biol 29:1–8

Halligan JJ et al (2016) Pre-Clovis occupation 14,550 years ago at the Page-Ladson site, Florida, and the peopling of the Americas. Sci Adv 2:e1600375

Han B-C (2014) Psicopolítica. Herder Editorial, Barcelona

Harari YN (2015) Sapiens. A brief history of humankind, Vintage, London

Harari YN (2016) Homo deus. A brief history of tomorrow, Vintage, London

Hardi K et al (2012) Neanderthal medics? Evidence for food, cooking, and medicinal plants entrapped in dental calculus. Naturwissenschaften 99:617–626

Hare B (2017), Survival of the friendliest: *Homo sapiens* evolved via selection for prosociality. Annual Review of Psychology 68:155–186

Hare B et al (2012) The self-domestication hypothesis: evolution of bonobo psychology is due to selection against aggression. Anim Behav 83:573–585

Harmand S et al (2015) 3.3-Milion-year-old stone tools from Lomekwi 3, West Turkana, Kenya. Nature 521:310–315

Harvati K et al (2019) Apidina cave fossils provide earliest evidence of *Homo sapiens* in Eurasia. Nature 571:500–504

Hawks J (2016). https://aeon.co/ideas/human-evolution-is-more-a-muddy-delta-than-a-branching-tree

Hawks J et al (2017) New fossil remains of *Homo nal*edi from the Lesedi Chamber, South Africa. eLife. https://doi.org/10.7554/eLife.24232

Henneberg M et al (2014) Evolved developmental homeostasis disturbed in LB1 from Flores, Indonesia, denotes Down syndrome and not diagnostic traits of the invalid species *Homo floresiensis*. Proc Natl Acad Sci U S A 111(33):11967–11972. pmid:25092311

Henry AG et al (2010) Microfossils in calculus demonstrate consumption of plants and cooked foods in Neanderthal diets (Shanidar III, Iraq; Spy I and II, Belgium). Proc Natl Acad Sci 108:486–491

Henry AG et al (2012) The diet of *Australopithecus Sediba*. Nature 487:90–93

Henshilwood CS et al (2002) Emergence of modern human behaviour: middle stone age engravings from South Africa. Science 295:1278–1280

Henshilwood CS et al (2004) Middle stone age Shell beads from South Africa. Science 304:404

Henshilwood CS et al (2009) Engraved Ochres from the middle stone age levels at Blombos Cave, South Africa. J Hum Evo 57:27–47

Hershkovitz I et al (2018) Levantine cranium from Manot Cave (Israel) foreshadows the first European modern humans. Nature 520:216–219

Higham T et al (2014) The timing and spatiotemporal patterning of Neanderthal disappearance. Nature 512:306–309

Hlubik S et al (2016) Hominin fire use in the Okote member at Koobi Fora, Kenya:new evidence for the old debate. J Hum Evol 133:214–229

Hobaiter C, Byrne RW (2014) The meanings of Chimpanzee Gestures. Curr Biol. https://doi.org/10.1016/j.cub.2014.05.066

Hodgskiss T, Wadley L (2017) How people used ochre at Rose Cottage Cave, South Africa: sixty thousand years of evidence from the Middle Stone Age. PLoS One 12(4):e0176317. https://doi.org/10.1371/journal.pone.0176317

Hoffman BU, Lumpkin EA (2018) A gut feeling. Science 361:1203–1204

Hoffmann DL et al (2018a) U-Th dating of carbonate crusts reveals Neandertal origin of Iberian cave art. Science 359:912–915

Hoffmann DL et al (2018b) Symbolic use of marine shells and mineral pigments by Iberian Neandertals 115,000 years ago. Sci Adv 4:eaar5255

Holloway RL et al (2018) Endocast morphology of Homo naledi from the Dinaledi Chamber, South Africa. Proc Nat Acad Sci 115:5738–5743

Hublin J-J (2017) New fossils from Jebel Irhoud, Morocco, and the pan-African origin of *Homo sapiens*. Nature 546:289–292

Huerta-Sánchez E (2014) Altitude adaptation in Tibetans caused by introgression of Denisovan-like DNA. Nature 512:194–197

Iakovleva l et al (2012) The late Upper Palaeolithic site of Gontsy (Ukraine): a reference for the reconstruction of the hunter-gatherer system based on a mammoth. Quat Int 255:86–93

Jablonski NG et al (2010) Human skin pigmentation as an adaptation to UV radiation. Proc Acad Sci 107:8962–8968

Japyassú HF, Laland KN (2017) Extended spider cognition. Anim Cog 20:375–395

Jones RS (2015) Space diet: daily mealworm (*Tenebrio molitor*) harvest on a multigenerational spaceship. J Interdisc Sci Top 4:1–4

Joordens JCA et al (2015) *Homo erectus* at Trinil on Java used shells for tool production and engraving. Nature 518:228–231

Kano F, Hirata S (2015) Great Apes make anticipatory looks based on long-term memory of single events. Curr Biol 25:2513–2517

Kirby S (2017) Culture and biology in the origins of linguistic structure. Psyhonomic Bull Rev 24:118–137

Kittler R et al (2003) Molecular evolution of pedinculus humanus and the origin of clothing. Curr Biol 13:1414–1417

Krause J et al (2010) The complete mitochondrial DNA genome of an unknown Hominin from Southern Siberia. Nature 464:894–897

Kubota JT et al (2012) The neuroscience of race. Nat Neurosci 15(7):940–948

Kuhlwilm M et al (2016) Ancient gene flow from early modern humans into Eastern Neanderthals. Nature. https://doi.org/10.1038/nature16544

Laland KN (2017) Darwin's unfinished simphony. How culture made the human mind. Princeton University Press, Princeton and Oxford

Lalueza-Fox C et al (2007) A Melancortin 1 receptor allele suggests varying pigmentation among Neanderthals. Science 318:1453–1455

Lalueza-Fox C et al (2009) Bitter taste perception in Neanderthals through the analysis of the TAS2R38 gene. Biol Lett 5:809–811

Lambert C (2017) Il lavoro ombra. Baldini & Castoldi, Milano

Leach HM (2003) Human domestication reconsidered. Curr Anthropol 44:349–368

Leach HM (2007) Selection and the unforeseen consequences of domestication. In: Where the wild things are now: domestication reconsidered. Berg, Oxford, NY

Lebel S, Trinkaus E (2001) A carious Neanderthal Molar from the Bau de l'Aubésier, Vaucluse. France J Archaeol Sci 29:555–557

Ledogar JA et al (2016) Mechanical evidence that *Australopithecus sediba* was limited in its ability to eat hard foods. Nat Commun. https://doi.org/10.1038/ncomms10596

Ledoux J (1998) The emotional brain. The mysterious underpinning of emotional life. Simon & Schuster, NY

Leibo, J.Z. et al. (2017) Multi-agent reinforcement learning in sequential social dilemmas. In: S. Das, E. Durfee, K. Larson, M. Winiko (Eds.), Proceedings of the 16th international conference on autonomous agents and multiagent systems (AAMAS 2017), May 8–12, 2017, São Paulo, Brazil

Lewis, M. et al. (2017) Deal or no deal? Training AI bots to negotiate. Artif Intell Res. https://code.facebook.com/posts/1686672014972296/deal-or-no-deal-training-ai-bots-to-negotiate

Liu W et al (2010) Human remains from Zhirendong, South China, and modern human emergencein East Asia. Proc Natl Acad Sci 107:19201–19206

Liu W et al (2015) The earliest unequivocally modern humans in southern China. Nature 526:696–699

Loverdidge AJ et al (2016) Conservation of large predator populations: demographic and spatial responses of African lions to the intensity of trophy hunting. Biol Conserv 104:247–254

Manzi G (2016) Humans of the Middle Pleistocene: the controversial calvarium from Ceprano (Italy) and its significance for the origin and variability of *Homo heidelbergensis*. Quat Int 411:254–261

Mar AM (2011) The neural bases of social cognition and story comprehension. Annu Rev Psychol 62:103–134

Marean CW (2015) An evolutionary anthropological perspective on modern human origins. Annu Rev Anthropol 44

Marean CW (2017) The transition to foraging for dense an predictable resources and its impact on the evolution of modern humans. Phil Trans R Soc B 371:20150239. https://doi.org/10.1098/rstb.2015.0239

Marsh DE (2007) he origins of diversity: Darwin's conditions and epigenetic variations. Nutr Health 19:103–132

Marshall A (1890) Principles of economics. McMillan, London

Martin F (2014) Money: an unauthorized biography. Knopf, New York

Martinon-Torres M et al (2018) A "source and sink" model for East Asia? Preliminary approach through the dental evidence. Comptes Rendus Palevol 17:33–43

Mcbrearty S, Brook AS (2000) The revolution that wasn't: a new interpretation of the origin of modern human behavior. J Hum Evol 39:453–563

McPherron SP et al (2010) Evidence for Stone-tool-assisted consumption of animal tissues before 3.39 million years ago at Dikika, Ethiopia. Nature 466:857–860

Mellars P, French JC (2011) Tenfold population increase in Western Europe at the Neanderthal-to-modern human transition. Science 333:623–627

Meyer M et al (2012) A high-coverage genome sequence from an archaic Denisovan individual. Science 338:222–226

Miller G (2000) The mating mind: how sexual choice shaped the evolution of human nature. Doubleday, New York

Miller GH et al (1999) Pleistocene extinction of *Genyornis newtoni*: human impact on Australian megafauna. Science 8:205–208

Miller GH et al (2015) Human predation contributed to the extinction of the Australian megafaunal bird Genyornis newtoni ~47 ka. Nat Commun. https://doi.org/10.1038/ncomms10496

Mills KL et al (2014) The developmental mismatch in structural brain maturation during adolescence. Dev Neurosci 36:147–160

Mirazón Lahr M et al (2016) Inter-group violence among early holocene hunter-gatherers of West Turkana, Kenya. Nature. https://doi.org/10.1038/nature16477

Mitteroecker P et al (2016) Cliff-edge model of obstetric selection in humans. Proc Natl Acad Sci 113:14680–14685
Monge J et al (2013) Fibrous dysplasia in a 120,000+ year old Neandertal from Krapina, Croatia. PLoS One 8(6):e64539. https://doi.org/10.1371/journal.pone.0064539
Morgan E (1972) The descent of woman. Souvenir Press, London
Morris D (1967) The Naked Ape: a zoologist's study of the human animal. Jonatham Cape Publishing, London
Morwood MJ et al (2004) Archaeology and age of a new hominin from Flores in eastern Indonesia. Nature 431:1087–1091
Morwood MJ et al (2005) Further evidence for small-bodied hominins from the Late Pleistocene of Flores, Indonesia. Nature 437:1012–1017
Motesharrei S et al (2016) Modeling sustainability: population, inequality, consumption, and bidirectional coupling of the Earth and human systems. Natl Sci Rev 3:470–494
Mounk Y (2018) The people vs democracy why our freedom is in danger & how to save it. Harvard University Press
Muthukrishna M et al (2018) The cultural brain hypothesis: how culture drives brain expansion, sociality, and life history. PLoS Comput Biol 14(11):e1006504. https://doi.org/10.1371/journal.pcbi.1006504
Nava A et al (2017) Virtual histological assessment of the prenatal life history and age at death of the Upper Paleolithic fetus from Ostuni (Italy). Sci Rep 7:9427
Neubauer S, Hublin J-J, Gunz P (2018) The evolution of modern human shape. Sci Adv 4:eaao5961
Nowell A, Pettitt P (2012) The palaeolithic origins of human burial. Camb Archaeol J 22:298–299
Ottoni C et al (2017) The palaeogenetics of cat dispersal in the ancient world. Nat Ecol Evol. https://doi.org/10.1038/s41559-017-0139
Oxilia G et al (2015) Earliest evidence of dental caries manipulation in the Late Upper Palaeolithic. Nat Scientific Rep. https://doi.org/10.1038/srep12150
Pagani L et al (2016) Genomic analyses inform on migration events during the peopling of Eurasia. Nature 538:238–242
Page AE et al (2016) Reproductive trade-offs in extant hunter-gatherers suggest adaptive mechanism for the Neolithic expansion. Proc Natl Acad Sci 113:4694–4699
Pascanu R et al (2017) Learning model-based planning from scratch, arXiv:1707.06170v1 [cs.AI]
Pearce E et al (2013) New insights into differences in brain organization between Neanderthals and anatomically modern humans. Proc R Soc B Biol Sci 280:1–7
Penfield W, Boldrey E (1937) Somatic motor and sensory representation in the cerebral cortex of man as studied by electrical stimulation. Brain 60:389–440
Peresani M et al (2011) Late Neandertals and the intentional removal of feathers as evidenced from bird bone taphonomy at fumane cave 44 ky B.P., Italy. Proc Natl Acad Sci 108:3888–3893
Perner J, Esken F (2015) Evolution of human cooperation in Homo heidelbergensis: teleology versus mentalism. Dev Rev 28:69–88
Perry GLW et al (2012) Explaining fire-driven landscape transformation during the Initial Burning Period of New Zealand's prehistory. Glob Chang Biol 18:1609–1621
Pettitt PB (2010) The palaeolithic origins of human Burial. Routledge, London
Pettitt PB et al (2003) The Gravettian Burial known as the Prince ("Il Principe"): new evidence for his age and diet. Antiquity 77:15–19
Pfeifer M et al (2017) Creation of forest edges has a global impact on forest vertebrates. Nature 551:187–119
Pike AWG et al (2012) U-Series dating of paleolithic art in 11 caves in Spain. Science 336:1409–1413
Pinker S, Jackendoff R (2005) The faculty of language: what's special about it? In: Cognition, vol 95, pp 201–236
Pinkola Estés C (1992) Women who run with the wolves, Rider, UK

Posth C et al (2017) Deeply divergent archaic mitochondrial genome provides lower time boundary for African gene flow into Neanderthals. Nat Commun 8:16046. https://doi.org/10.1038/ncomms16046. www.nature.com/naturecommunications

Potter BA et al (2014) New insights into Eastern Beringian mortuary behavior: a terminal Pleistocene double infant burial at Upward Sun River. Proc Natl Acad Sci 111:17060–17065

Powell A et al (2009) Late Pleistocene demography and the appearance of modern human behaviour. Science 324:1298–1301

Powell J et al (2012) Orbital prefrontal cortex volume predicts social network size: an imaging study of individual differences in humans. Proc R Soc B: Biol Sci 279:2157–2162

Prüfer K et al (2012) The bonobo genome compared with the chimpanzee and human genomes. Nature 486:527–531

Prüfer K et al (2014) The complete genome sequence of a Neanderthal from the Altai Mountains. Nature 505:43–49

Prüfer K et al (2017) A high-coverage Neandertal genome from Vindija Cave in Croatia. Science 358:655–658

Purzycki BG et al (2016) Moralistic gods, supernatural punishment and the expansion of human sociality. Nature 530:327–330

Putt SS et al (2017) The functional brain networks that underlie Early Stone Age tool manufacture. Nat Hum Behav. https://doi.org/10.1038/s41562-017-0102

Quiles A et al (2015) A high-precision chronological model for the decorated upper Paleolithic cave of Chauvet-Port d'Arc, Ardèche, France. Proc Natl Acad Sci 113:4670–4675

Radovčić D et al (2015) Evidence for Neandertal jewelry: modified white-tailed eagle claws at Krapina. PLoS One 10(3):e0119802. https://doi.org/10.1371/journal.pone.0119802

Raghavan M (2014) Upper Palaeolithic Siberian genome reveals dual ancestry of Native Americans. Nature 505:87–91

Rasmussen M et al (2011) An aboriginal Australian genome reveals separate human dispersals into Asia. Science 334:94–98

Rasmussen M et al (2014) The genome of a Late Pleistocene human from a Clovis burial site in western Montana. Nature 506:225–229

Raymo ME, Huybers P (2008) Unlocking the mysteries of the ice ages. Nature 451:284–285

Reich D (2018) Who we are and how we got here. Ancient DNA and the new science of the human past. Pantheon Books, New York

Reich D et al (2010) Genetic history of an archaic hominin group from Denisova Cave in Siberia. Nature 468:1053–1060

Rendu W et al (2014) Evidence supporting an intentional Neanderthal Burial at La Chappelle-aux-Saints. Proc Natl Acad Sci 111:81–86

Richter D et al (2017) The age of hominin fossils from Jebel Irhoud, Morocco, and the origins of the Middle Stone Age. Nature 546:293–296

Riel-Salvatore J, Gravel-Miguel C (2013) Upper palaeolithic mortuary practice in Eurasia: a critical look at the burial record. In: Tarlow S, Nilsson Stutz L (eds) The Oxford handbook of the archaeology of death and Burial. Oxford University Press, Oxford, pp 303–346

Rifkin RF et al (2015) Evaluating the photoprotective effects of Ochre on human skin by in vivo SPF assessment: implications for human evolution, adaptation and dispersal. PLoS One 10(9): e0136090. https://doi.org/10.1371/journal.pone.0136090

Rizzolatti G, Craighero L (2004) The mirror-neuron system. Annu Rev Neurosci 27:169–192

Rizzolatti R, Sinigaglia C (2016) The mirror mechanism: a basic principle of brain function. Nat Rev/Neurosci. https://doi.org/10.1038/nrn.2016.135

Roach NT (2013) Elastic energy storage in the shoulder and the evolution of high-speed throwing in Homo. Nature 498:483–486

Roberts A, Maslin M (2016) Sorry David Attenborough, we didn't evolve from 'aquatic apes' – here's why. The Conversation

Roberts LG et al (2001) New ages for the last Australian Megafauna: continent- wide extinction about 46,000 Years ago. Science 292:1888–1892

References

Roberts LG et al (2016) Climate change not to blame for late quaternary megafauna extinctions in Australia. Nat Commun. https://doi.org/10.1038/ncomms10511

Rodríguez-Vidal J et al (2014) A rock engraving made by Neanderthals in Gibraltar. Proc Natl Acad Sci 111:13301–13306

Romandini N et al (2014) Convergent evidence of Eagle Talons used by late Neanderthals in Europe: a further assessment on symbolism. Proc Acad Sci 9(7):e101278

Rosenberg K, Trevathan W (1995) Bipedalism and human birth: the obstetrical dilemma revisited. Evol Anthropol 4:161–168

Rougier H et al (2016) Neandertal cannibalism and Neandertal bones used as tools in Northern Europe. Nat Sci Rep 6. https://doi.org/10.1038/srep29005

Rovelli C (2018) The order of time 2018. Riverhead Boks, UK

Ruxton GD, Wilkinson DM (2011) Avoidance of overheating and selection for both hair loss and bipedality in hominins. Proc Natl Acad Sci 108:20965–20969

Salazar-López E et al (2012) The mental and subjective skin: emotion, empathy, feelings and thermography. Conscious Cogn 34:149–162

Sanchez G et al (2014) Human (Clovis)–gomphothere (Cuvieronius sp.) association ~13,390 calibrated yBP in Sonora, Mexico. Proc Nat Acad Sci 29:10972–10977

Sankararaman S et al (2014) The genomic landscape of Neanderthal Ancestry in present-day humans. Nature 507:354–357

Scerri EML et al (2018) Did our species evolve in subdivided populations across Africa, and why does it matter? Cell 33:582–594

Schlebush CM et al (2017) Southern African ancient genomes estimate modern human divergence to 350,000 to 260,000 years ago. Science 358:652–655

Schwartz JH, Tattersall I (2015) Defining the genus *Homo*. Science 349:931–932

Schwitalla AW et al (2014) Violence among Foragers: the bioarchaeological record from Central California. J Anthropol Archaeol 33:66–83

Shanley DP et al (2007) Testing evolutionary theories of menopause. Proc Royal Soc B: Biol Sci. https://doi.org/10.1098/rspb.2007.1028

Shipman P (2015) How do you kill 86 mammoths? Taphonomic investigations of mammoth megasite. Quater Int:359–360

Shipman P (2017) The invaders. How humans and their dogs Drove Neanderthals to extinction. Harvard University Press

Sikora M et al (2017) Ancient genomes show social and reproductive behavior of early Upper Paleolithic foragers. Science 358:659–662

Silk JB et al (2013) Chimpanzees share food for many reasons: the role of kinship, reciprocity, social bonds and harassment on food transfers. Anim Behav 85:941–947

Silver D et al (2016) Mastering the game of Go with deep neural networks and tree search. Nature 529:484–489

Silver D et al (2017) Mastering the game of Go without human knowledge. Nature 550:354–359

Silver D et al (2018) A general reinforcement learning algorithm that masters chess, shogi, and go through self-play. Science 362:1140–1144

Simonti CN et al (2016) The phenotypic legacy of admixture between modern humans and Neandertals. Science 351:737–741

Simpson SW et al (2008) A female *Homo Erectus* pelvis from Gona. Ethiopia Sci 322:1089–1092

Sistiaga A et al (2014) The Neanderthal meal: a new perspective using faecal biomarker. PLoS One 9(6):e101045. https://doi.org/10.1371/journal.pone.0101045

Slaughter A-M (2017) The chessboard and the Web: strategies of connection in a networked world. Yale University Press, New Haven (CT)

Sloterdjik P (2018) What happened in the 20th century. Politi Press, Cambridge, UK

Smail DL (2007) On deep history and the brain. University of California Press, Berkeley

Smith A (1776) The wealth of nations. W. Strahan and T. Cadell, London.

Smith GM (2014) Neanderthal megafaunal exploitation in Western Europe and its dietary implications: a contextual reassessment of La Cotte de St Brelade (Jersey). J Hum Evol 78:181–201

Smith TM (2018) The tales teeth tell. The MIT Press, Cambridge, MA

Smith TM et al (2007) Earliest evidence of modern human life history in North African early *Homo sapiens*. Proc Natl Acad Sci 104:6128–6133

Smith KF et al (2014) Global rise in human infectious disease outbreaks. J R Soc Interface 11:20140950. https://doi.org/10.1098/rsif.2014.0950

Smith TM et al (2015) Dental ontogeny in pliocene and early pleistocene hominins. PLoS One. https://doi.org/10.1371/journal.pone.0118118

Soressi M et al (2013) Neanderthals made the first specialized bone tools in Europe. Proc Natl Acad Sci 110:14186–14190

Sponheimer M, Lee-Thorp J (1999) Isotopic evidence for the diet of an Early Hominid, *Australopithecus africanus*. Science 283:368–370

Sponheimer M et al (2013) Isotopic evidence of early hominins diets. Proc Natl Acad Sci U S A 110:10513–10518

Stedman HH et al (2004) Myosin gene mutation correlates with anatomical changes in the human lineage. Nature 428:415–418

Stilwell F (2019) The political economy of inequality. Polity Press, Cambridge

Stiner MC et al (2009) Cooperative hunting and meat sharing 400–200 kya at Qesem Cave, Israel. Proc Natl Acad Sci 106:207–213

Stout D, Hecht E (2015) Neuroarchaeology. In: Bruner E (ed) Hum Palaeoneurol. Springer series in Bio-/Neuroinformatics, pp. 145–175

Stout D et al (2008) Neural correlates of Early Stone Age toolmaking: technology, language and cognition in human evolution. Phil Trans R Soc B363:1939–1949

Stout D et al (2017) Evolutionary neuroscience of cumulative culture. Proc Nat Acad Sci 114:7861–7868

Stringer CB (2016) The origin and evolution of *Homo sapiens*. Phil Trans R Soc B 371:20150237. https://doi.org/10.1098/rstb.2015.0237

Stringer CB (2019) The new human story. *Financial Times*, 26 July

Stringer CB, Andrews P (1988) Genetic and fossil evidencefor the origin of modern humans. Science 239:1263–1268

Suddendorf T (2013) The gap: the science of what separates us from other animals. Basic Books

Suddendorf T et al (2018) Prospection and natural selection. Curr Opin Behav Sci 24:26–31

Sutikna T et al (2016) Revised stratigraphy and chronology for *Homo floresiensis* at Liang Bua in Indonesia. Nature 532:366–369

Sutou S (2012) Hairless mutation: a driving force of humanization from a human-ape common ancestor by enforcing upright walking while holding a baby with both hands. Genes Cells 17:264–272

Tainter JA (2003) The collapse of complex societies. Cambridge University Press, Cambridge

Takemoto H (2017) Acquisition of terrestrial life by human ancestors influenced by forest microclimate. Nat Scientific Rep 7:5741. https://doi.org/10.1038/s41598-017-05942-5

Takemoto H et al (2015) How did Bonobos come to range South of the Congo River? Reconsideration of the divergence of *Pan paniscus* from other pan populations. Evol Anthropol 24:170–184

Tattersall I (2015) The strange case of the Rickety Cossack. Palgrave Macmillan, New York

Tattersall I (2017) Why was human evolution so rapid? In: Marom A, Hovers E (eds) Human paleontology and prehistory, vertebrate paleobiology and paleoanthropology. https://doi.org/10.1007/978-3-319-46646-0_1

Terradas J (2017) Towards a general theory of evolution. Blog.creaf.cat/en/coneixement/towards-a-general-theory-of-evolution/

Thaler RH (2015) Misbehaving: the making of behavioural economics. W.W. Norton & Company, New York

Theofanopoulou C et al (2017) Self-domestication in *Homo sapiens*: insights from comparative genomics. PLoS One 12(10):e0185306. https://doi.org/10.1371/journal.pone.0185306

Thomas J, Kirby S (2018) Self domestication and the evolution of language. Biol Phylos 33(1):9. https://doi.org/10.1007/s10539-018-9612-8
Tiberi Vipraio P (1999) Dal Mercantilismo alla Globalizzazione. Lo Sviluppo Industriale Trainato dalle Esportazioni, il Mulino, Bologna
Tierney JE et al (2017) A climatic context for the out-of-Africa migration. Geology. https://doi.org/10.1130/G39457.1
Timmermann A, Friedrich T (2016) Late Pleistocene climate drivers of early human migration. Nature 538:92–95
Tobias PV (1997) Il bipede barcollante. Einaudi, Torino
Tobler R et al (2017) Aboriginal mitogenomes reveal 50,000 years of regionalism in Australia. Nature 544:180–184
Tomasello M (2016) A natural history of human morality. Harvard University Press
Toups MA et al (2011) Origin of clothing Lice indicates Early clothing use by anatomically modern humans in Africa. Mol Biol Evol 28:29–32
Trinkaus E (1983) The Shanidar Neandertals. Academic, New York
Trinkaus E, Shang H (2008) Anatomical evidence for the antiquity of human footwear: Tianyuan and Sunghir. J Archaeol Sci 35:1928–1933
Trut L (2001) Experimental studies of early canid domestication. In: Ruvinsky A, Sampson J (eds) The genetics of the dog. CABI, New York, pp 15–42
Trut L et al (2009) Animal evolution during domestication: the domesticated fox as a model. BioEssays 31:349–360
Tuniz C (2012) Radioactivity: a very short introduction. Oxford University Press, Oxford
Tuniz C, Tiberi Vipraio P (2016) Humans. An unauthorized biography, Springer/Nature, Heidelberg
Tuniz C, Tiberi Vipraio P (2018) La Scimmia Vestita. Dalle Tribù di Primati all'Intelligenza Artificiale. Carocci editore, Roma
Tuniz C, Gillespie R, Jones C (2009) The bone readers. Routledge, USA
Tuniz C, Manzi G, Caramelli D (2014) The science of human origins. Routledge, USA
Turchin P (2003) Historical dynamics: why states rise and fall. Princeton University Press, Princeton
Turner F (2008) From counterculture to cyberculture. University of Chicago Press, Chicago
Turney CSM et al (2001) Redating the onset of burning Lynch's Crater (North Queensland): implications for human settlement in Australia. J Quat Sci 16:767–771
Upadhayay N et al (2014) Comparison of cognitive functions between male and female medical students: a pilot study. J Clin Diagn Res. 8(6):BC12–BC15. https://doi.org/10.7860/JCDR/2014/7490.4449
van den Bergh GD et al (2016a) Earliest hominin occupation of Sulawesi, Indonesia. Nature 529:208–211
van den Bergh G et al (2016b) *Homo floresiensis*-like fossils from the early Middle Pleistocene of Flores. Nature 534:245–248
Vanhaeren M, d'Errico F (2001) Personal Ornaments from the La Madeleine child (Peyrony excavations): an insight into upper palaeolithic childhood. Revd'Archeol Préhist 13:201–240
Vanhaeren M, d'Errico F (2005) Grave goods from the Saint-Germain-la-Rivière Burial: evidence for Social Inequality in the Upper Palaeolithic. J Anthropol Archaeol 24:117–134
Vanhaeren M et al (2013) Thinking strings: additional evidence for personal ornament use in the middle stone age of Blombos Cave. South Africa J Hum Evol 64:500–517
Villmoare B et al (2015) Early *Homo* at 2.8 Ma from Ledi-Geraru, Afar, Ethiopia. Science 347:1352–1355
Vitali S et al (2011) The network of global corporate control. PLoS One 6(10):e25995. https://doi.org/10.1371/journal.pone.0025995
Walter RC (1994) Age of lucy and the first family: single-crystal 40Ar/39Ar dating of the Denen Dora and Lower Kada Hadar Members of the Hadar Formation, Ethiopia. Geology 22:6–10

Waters CN et al (2016) The Anthropocene is functionally and stratigraphically distinct from the Holocene. Science 351. https://doi.org/10.1126/science.aad2622

Watts I (2009) Red ochre, body-painting, and language: interpreting the Blombos ochre. In: Botha R, Knight C (eds) The cradle of language, vol 2. Oxford University Press, Oxford, pp 93–129

Weber T et al (2017) Imagination-augmented agents for deep reinforcement learning, arXiv:1707.06203v1

Weiss A et al (2012) Evidence for a midlife crisis in great apes consistent with the U-shape in human well-being. Proc Natl Acad Sci 109:19949–19952

Weiss MC et al (2016) The physiology and habitat of the last universal common ancestor. Nat Microbiol. https://doi.org/10.1038/nmicrobiol.2016.116

Westaway et al (2017) An early modern human presence in Sumatra 73,000–63,000 years ago. Nature. https://doi.org/10.1038/nature23452

Weyrich LS et al (2017) Neanderthal behaviour, diet, and disease inferred from ancient DNA in dental calculus. Nature. https://doi.org/10.1038/nature21674

Wheeler PE (1984) The evolution of bipedality and loss of functional body hair in hominids. J Hum Evol 13:91–98

White TD et al (2009) *Ardipithecus ramidus* and the Paleobiology of Early Hominids. Science 326:75–86

White TD et al (2015) Neither chimpanzee nor human, *Ardipithecus* reveals the surprising ancestry of both. Proc Natl Acad Sci 112:4877–4884

Wiener N (1948) Cybernetics or control and communication in the animal and the machine. Technology Press and John Wiley & Sons, Cambridge, MA

Wilkins AS, Wrangham R W, Fitch W T (2014) The domestication syndrome. In: Mammals: a unified explanation based on neural crest cell behavior and genetics genetics, vol. 197, pp. 795–808

Wilkins J et al (2017) Lithic technological responses to Late Pleistocene glacial cycling at Pinnacle Point Site 5-6, South Africa. PLoS One. https://doi.org/10.1371/journal.pone.0174051

Williams AN (2013) A new population curve for prehistoric Australia. Proc R Soc Lond B 280:20130486. https://doi.org/10.1098/rspb.2013.0486

Wolpoff MH, Caspari R (1997) Race and human evolution. Simon and Schuster, London

Wrangham RW (2014) Did *Homo sapiens* self-domesticate? Presented at the center for Academic Research and training in anthropogeny: domestication and human evolution. La Jolla, CA. Retrieved from https://www.youtube.com/watch?v=acOZT240bTA

Wrangham RW (2018) Two types of aggression in human evolution. Proc Natl Acad Sci U S A 115:245–253

Wrangham RW (2019) The goodness paradox. How evolution made us both more and less violent. Profile Books, United Kingdom

Wynn T (1995) Handaxe enigmas. World Archaeol 27:10–24

Xin J et al (2019) Brain differences between men and women: evidence from deep learning. Front Neurosci 13:185. https://doi.org/10.3389/fnins.2019.00185

Xu D et al (2017) Archaic Hominin introgression in Africa contributes to functional salivary MUC7 genetic variation. Mol Biol Evol. https://doi.org/10.1093/molbev/msx206

Yamins DLK, Di Carlo JJ (2016) Using goal-driven deep learning models to understand sensory cortex. Nat Neurosci 19:356–336

Yang et al (2016) Tending a dying adult in a wild multi-level primate society. Curr Biol. https://doi.org/10.1016/j.cub.2016.03.062

Zahid HJ et al (2016) Agriculture, population growth, and statistical analysis of the radiocarbon record. Proc Natil Acade Sci 113:931–935

Zanolli C, Mazurier A (2013) Endostructural characterization of the *H. Heidelbergensis* dental remains from the Early Middle Pleistocene Site of Tighenif, Algeria. Comptes Rendus Palevol 12:293–304

Zilhão J et al (2010) Symbolic use of marine shells and mineral pigments by Iberian Neandertals. Proc Natl Acad Sci 107:1023–1028

Zipfel B, Berger LR (2007) Shod versus unshod: the emergence of forefoot pathology in modern humans? Foot 17:205–213

Made in the USA
Las Vegas, NV
06 March 2025